Collaborative Research Methods in the Arctic

This book addresses the growing demand for collaborative and reflexive scholarly engagement in the Arctic directed at providing relevant insights to tackle local challenges of arctic communities. It examines how Arctic research can come to matter in new ways by combining methods and engagement in the field of inquiry in new and meaningful ways.

Research informs decisions affecting the futures of Arctic communities. Owing to its ability to include local concerns and practices, collaborative research could play a greater role in this process. By way of example of how to bring new voices to the fore in research, this edited collection presents experiences of researchers active in collaborative Arctic research. It draws on multidisciplinary perspectives from a broad range of academics in the fields such as law and medicine over tourism and business studies, planning and development, cultural studies, ethnology, and anthropology. It also shares personal experiences of working in Greenland and with Greenlanders, whether communities, businesses and entrepreneurs, public officials and planners, patients, or students.

Offering useful insights into the current problems of Greenland that are representative of the Arctic region, this book will be beneficial for researchers and scientists involved in Arctic research.

Anne Merrild Hansen is professor of planning and impact assessment in the Arctic at Aalborg University and the head of the Research Platform AAU Arctic at Aalborg University. Her research is focused on social impact assessment in relation to extractive industries in the Arctic.

Carina Ren is associate professor of tourism and cultural innovation at the Centre for Innovation and Research in Culture and Living in the Arctic at Aalborg University and the platform coordinator of AAU Arctic. Carina researches connections between tourism and other fields of society through cultural innovation, knowledge collaboration, and capacity building.

Routledge Research in Polar Regions

Series Editor: Timothy Heleniak, *Nordregio International Research Centre, Sweden*

The Routledge series in Polar Regions seeks to include research and policy debates about trends and events taking place in two important world regions: the Arctic and Antarctic. Previously neglected periphery regions, with climate change, resource development and shifting geopolitics, these regions are becoming increasingly crucial to happenings outside these regions. At the same time, the economies, societies and natural environments of the Arctic are undergoing rapid change. This series seeks to draw upon fieldwork, satellite observations, archival studies and other research methods which inform about crucial developments in the Polar regions. It is interdisciplinary, drawing on the work from the social sciences and humanities, bringing together cutting-edge research in the Polar regions with the policy implications.

For more information about this series, please visit: www.routledge.com/Routledge-Research-in-Polar-Regions/book-series/RRPS

Collaborative Research Methods in the Arctic

Experiences from Greenland

Edited by Anne Merrild Hansen and Carina Ren

Routledge
Taylor & Francis Group

LONDON AND NEW YORK

First published 2021
by Routledge
2 Park Square, Milton Park, Abingdon, Oxon OX14 4RN

and by Routledge
605 Third Avenue, New York, NY 10017

First issued in paperback 2022

Routledge is an imprint of the Taylor & Francis Group, an informa business

Publisher's Note
The publisher has gone to great lengths to ensure the quality of this reprint but points out that some imperfections in the original copies may be apparent.

British Library Cataloguing-in-Publication Data
A catalogue record for this book is available from the British Library

Library of Congress Cataloging-in-Publication Data
A catalog record has been requested for this book

ISBN: 978-0-367-46755-5 (hbk)
ISBN: 978-0-367-54649-6 (pbk)
ISBN: 978-1-003-03084-3 (ebk)

DOI: 10.4324/9781003030843

Typeset in Times New Roman
by Taylor & Francis Books

Contents

vi *Contents*

Illustrations

Figures

Tables

Contributors

Stig Andersen, MD, PhD, is clinical professor of Geriatric and Internal Medicine at Aalborg University (Denmark), Lead of the Master's Program in Medicine at Aalborg University (Denmark), professor of Arctic Medicine at Ilisimatusarfik, University of Greenland and the founder and Head of Arctic Health Research Centre, Aalborg University Hospital. Research areas covered in the Arctic include blood pressure and ischemic heart disease, nutrition and cold adaptation, iodine intake and thyroid function, vitamin D and skeletal health, hepatitis and liver, inflammation and persistent organic pollutants, body build and obesity, and disease patterns and functional limitations in old age in the Arctic. Results have been presented at Arctic and international meetings and conferences, and published in peer reviewed international journals, in addition to conference papers and international book chapters.

Daniela Chimirri is a PhD fellow at the Center for Innovation and Research in Culture and Living in the Arctic (CIRCLA) at Aalborg University. Her research centers on community collaboration for tourism development in Greenland.

Ina Drejer, PhD, is associate professor of innovation studies with a particular emphasis on impact and policy relevant analyses at the Department of Business and Management at Aalborg University. Ina's research interests include innovation and industrial dynamics, innovation and industrial development policy, university–industry interaction, inter-firm relations, regional economics, and place-based development. Her work on the Arctic centers on industrial development perspectives and challenges in Greenland.

Louise Faber, PhD in Law, was previously affiliated as associate professor at the Department of Law and the Department of Business and Management at Aalborg University. Louise's research areas are related to real estate. She deals with the legal framework for real estate (building and land), including purchase agreements, tenure, public land use and environmental law. She has also been involved with the establishment of a Greenlandic education of law under the project "North Atlantic Law Programme" at Aalborg University (2011–2017).

Allan Næs Gjerding, PhD, is associate professor of business administration and head of section at Department of Business and Management at Aalborg University. Allan's research focuses on interorganizational relationships, especially emphasizing organizational roles in ecosystems and submarkets, collaborative business models, and triple-helix activities. His current research interests are focused on industrial symbiosis and the use of blockchain in activity systems. His work on the Arctic centers on industrial development perspectives and challenges in Greenland.

Naja Dyrendom Graugaard. Inspired by her own mixed background in Denmark and Greenland, Naja's research involves Inuit sealing, Indigenous knowledge, alternative narratives, and coloniality and decolonization in the Arctic. She holds an MA in environmental science from York University (UK), a BA from Trent University (Canada), and is currently finishing her PhD studies at Aalborg University. Naja is also a mother, script writer, and theatre artist.

Anne Merrild Hansen, PhD, is professor at the Danish Centre for Environmental Assessment (DCEA) and head of the cross-faculty platform, AAU Arctic at Aalborg University. She is also the head of the PhD school at Ilisimatusarfik, University of Greenland. The focus of Anne's research is on the social impacts and benefits related to resource development in Arctic communities. Anne also studies impact assessment processes with a particular focus on participation procedures and inclusion of indigenous knowledge and local knowledge in the assessments.

Verena Gisela Huppert is a PhD fellow at the Center for Innovation and Research in Culture and Living in the Arctic (CIRCLA) at Aalborg University. Her project is co-financed by Aalborg University, the municipality of Sermersooq, the Greenland Business Association, and Arctic Consensus and focuses on labor market issues in the Arctic, specifically on how to recruit and retain labor in Greenland. Verena holds an MA in Arctic Studies from Aalborg University (2016) and currently has her main residence in Nuuk, Greenland.

Sanne Vammen Larsen, PhD, is associate professor at the Danish Centre for Environmental Assessment (DCEA) at Aalborg University. Sanne's research is focused on impact assessment within a general framework of sustainable development, environmental planning, and decision-making processes. Specifically she is focused on the integration of climate change, risk, uncertainty, and social impacts in EIA and SEA and has worked mainly in a European and Arctic setting, including as a scholar in the Fulbright Arctic Initiative.

Carina Ren, PhD, is associate professor at the Centre for Innovation and Research in Culture and Living in the Arctic (CIRCLA) at Aalborg University and the platform coordinator of AAU Arctic. Carina researches connections between tourism and other fields of society through cultural innovation, knowledge collaboration, and capacity building. She currently works on tourism development in a Greenlandic and Arctic context and its implications on Arctic futures. Carina has published in leading Arctic and tourism journals and is the co-editor of several special issues and books, recently *Co-Creating Tourism Research. Towards Collaborative Ways of Knowing* (2017, Routledge) and *Theories of Practice in Tourism* (2018, Routledge).

Robert C. Thomsen, PhD, is Associate Professor at Aalborg University where, between 2012 and 2016, he was Head of the Centre for Innovation and Research in Culture and Learning in the Arctic (CIRCLA). He holds a PhD in nationalism studies and has published mostly on autonomist-nationalist movements in Canada and the North Atlantic, including the monograph *Nationalism in Stateless Nations* (2010). Arctic research interests include national identity-building and self-government in Greenland and Arctic Canada, pan-Inuit and pan-Arctic movements, and cultural heritage tourism.

Foreword

Henrik Halkier

When trying to imagine Arctic research, most people envision men in a freezing, desolate landscape together with dogs, high-tech equipment—and lots of snow and ice. Bearded scientists braving the cold to collect data that, when analyzed back in the lab in Europe or North America, will help us to understand processes related to climate change. Results will appear in high-impact international journals published in London or Amsterdam and be debated at international conferences in New York and Paris: a fly-in/fly-out mode of research that is essentially untouched by "Arctic humans," except perhaps in logistic support roles.

Obviously, there is more to Arctic research. Over the years, numerous studies have been undertaken in which people actually living in the Arctic play a key role (e.g., within anthropology, sociology, and cultural studies). But while the lives and conditions of Arctic residents have become research objects, external researchers have largely continued to set the agenda, collect data, analyze this data in their offices in Europe or North America, and eventually publish their findings with Routledge or Elsevier; still fly-in/fly-out, with the local communities subjected to what might be referred to as "the colonial gaze."

Reimagining Arctic research is one of the key ambitions of this book. By the diversity of their examples, the chapters demonstrate the potential of other, more collaborative, approaches to doing research in the Arctic.

The origins of this emerging "participatory turn" can be traced back to different, often coexisting, types of concerns. For some, ethical or political concerns are paramount: participatory methods are seen as a social good or moral imperative and a means of establishing a more equal distribution of roles between (external) researchers and Arctic communities. Either by involving local stakeholders in the research process or, at the very least, disseminating results to them in the form of, for example, policy advice, town hall meetings, or education. In some parts of the Arctic, such as Canada, government regulates access for researchers. In other parts, a perceived "climate of research distrust" reinforces whatever ethical concerns the individual researcher may have. And for European researchers and European research projects, the EU's increasing focus on stakeholder participation and research impact prompts new ways of conceiving research.

But reimagining Arctic research is also driven by an ambition to increase the scope and depths of the research. Put simply: collaborative research strategies can enable researchers to access knowledge that otherwise would have been out of reach. Engaging Arctic stakeholders in defining research projects and questions can help to build long-term community engagement and impact, and involving locals in carrying out studies can be a way not only to approach indigenous and local knowledge in a cross-cultural setting, but also fundamentally to enable more legitimate, relevant, and valuable research contributions and impacts in the Arctic.

Re-imagining Arctic research is no mean feat. Cooperation and trust must be established and maintained, and individual researchers depend on previous researchers in the sense that poor fly-in/fly-out experiences can contribute to a climate of distrust or indifference that could obstruct even the most noble ambitions of those who arrive after them. In order to advance participatory research strategies, two elements must therefore be in place. First, Arctic authorities and external research sponsors must focus on research ethics and impacts, such as codes of conduct and institutional platforms for citizen and stakeholder involvement. Second, Arctic researchers must be willing to engage in new methods that, if more rewarding, can still be more cumbersome and demanding than "the old ways."

This book encourages the re-imagination of Arctic research by documenting how collaborative methods can make a considerable contribution across and between disciplines; set a constructive and engaging tone; and thereby inspire other researchers to take up the challenges. Reimagining Arctic research must be a collaborative effort in more ways than one—something to look forward to!

1 Collaborative research methods in the Arctic: Why and how?

Anne Merrild Hansen & Carina Ren

The Arctic is undergoing social and environmental transformation as accelerating climate change and the effects thereof are having a profound impact on the living conditions in the region (AMAP, 2017; Carson and Petersen, 2016). Climate change and its higher-order effects are therefore considered as a major driver of Arctic change and are causing further derived impacts worldwide (Keskitalo and Southcott, 2014). The further opening of global trade since the end of the Cold War, various free trade agreements, and bilateral investment treaties have all contributed to making the Arctic increasingly accessible and opened up opportunities for commerce (Koivurova and Lesser, 2016). Resource demands, tourism, transport, fisheries, and other economic development are also driving significant change in the Arctic (Stephen, 2018). The message of an opening Arctic is currently being promoted strongly by actors across the world as the Arctic region becomes of increasing geopolitical strategic interest and attracts global interest more generally.

This outlined situation goes hand in hand with an interest in accessing and promoting knowledge about the region, its climate, its culture, mineral and other resources, sea, defense, search and rescue, governance systems, people, business landscapes, environments, and much more. This increasing interest has led to an explosion in the research activities and funding opportunities available in the Arctic.

Research-driven knowledge is informing decisions worldwide on actions and the prioritization of investments, strategies, attitudes, and approaches to the Arctic. While this interest might be seen as a positive development in learning more about the region, this knowledge is by no means "innocent." Those who define the research topics, methodologies, interpretations, and conclusions hold the power to influence decision-making (Hansen et al., 2016). The question then becomes: "Who gets to define the research methodologies that dictate the future of Arctic research?"

Together with the above scenarios of increasing interest in and the "will to know" about the Arctic, there is a movement in progress in local Arctic communities that is working pro-actively to define the rules of engagement with researchers and collaborative approaches. In some areas, living up to such rules has become a requirement to gain permission to carry out fieldwork (Saxinger and First Nation of Na-cho-Nyak Dun, 2018). This has become the case in Canada, for example, where the so-called policy of OCAP (Ownership, Control, Access, and Possession) has been developed as a framework for local and indigenous communities to manage relationships with researchers coming from outside the community (The First Nations Information Governance Centre, 2014). The literature on collaborative research dates back decades, also with respect to the Arctic regions (Iglič et al., 2017). However, the increasing interest in the region, together with the expectations from Arctic peoples to research involvement and decision-making on topics that will influence their living conditions and future, have created an awareness concerning local engagement and participation in research projects.

In Greenland, which is the central area of investigation in this book, frameworks such as OCAP do not (yet) apply. This might be explained by that which Frank Sejersen characterized in 2004 as one of the striking differences between Greenland and other Arctic communities in the application of local knowledge in research. With the best possibilities in place in Greenland, where an elaborate and extensive political system based on indigenous self-government has been developed, Sejersen pointed out how it as a striking paradox that local knowledge is so relatively overlooked. For instance, it was only in the late 1990s and with the integration of local knowledge in resource monitoring and management that the situation in Greenland changed from "simple, political rhetoric to serious dialogue between stakeholders" (2004: 33). Thus, according to Sejersen, Greenland in 2004 only seemed to be on the threshold to undertake discussions concerning issues of local involvement and responsibility in research and policymaking, which had been discussed in Alaska and Canada for at least three decades. Since then, as stressed in a new "toolkit" for "Community Based Participatory Research in Greenland" (Rink and Reimer, 2019), times are changing, and the need to include local knowledge and communities in all stages of research is increasingly pointed out by various political, academic, and community stakeholders in Greenland.

This book and its contributions are focussed on this context of accelerating transformations and dynamically evolving requirements, increasing interests, and growing concerns about how knowledge about the Arctic is produced and applied, locally as well as globally. By presenting examples rooted in the context of Greenland, we add to the recent literature on collaborative methods in research in collections such as *Community-Based Participatory Research* (Deeb-Sossa, 2019) and *Practising Community-Based Participatory Research* (MacKinnon, 2018). These volumes all underline how applying collaborative research methods and collaborating with local stakeholders not only ensures that the knowledge is locally anchored and consolidated, but also that it can contribute to the raising of competences allowing nations and peoples to benefit from research activities and knowledge co-production. In this edited volume, such contributions include projects stemming from community collaboration by addressing *locally identified* issues and promoting *community* social change. Other contributions draw on cases featuring a broader range of partners, ranging from the Greenland business community, civil society, municipalities, labor organizations, non-governmental organization (NGO), and the research community. By increasing the array of colla- borators, our book expands a more traditional view on local communities by describing and critically addressing research collaborations with many differ- ent sorts of actors in line with the edited book *Co-Creating Tourism Research: Towards Collaborative Ways of Knowing* (Ren et al., 2017) and *Participatory Action Research (pocket guide)* (Lawson et al., 2015).

Rather than providing "a new approach," we provide a curated selection of approaches from different scientific disciplines, which takes the needs seriously and experiments with new modes of co-production by bringing together, enhancing, and elaborating on research in the Arctic. All of the contributions work with local stakeholders and describe how collaborative methods were deployed to work with these particular groups in various ways and to varying degrees. The purpose of the contributions is to share and critically examine these experiences. Sharing our experiences allows for inspiration and discussion across disciplines concerning the future research needs in areas requiring fieldwork and engagement with local stakeholders, within the Arctic and beyond, while remaining attentive to the differences across the different Arctic communities. We hope that the book will provide food for thought and comparison with other researchers and research insti- tutions active in Arctic research to discuss appropriate and valuable ways of conducting research in the Arctic while at the same time bringing local voices to the forefront in research.

Situating Arctic research

This edited volume and the work behind it are born out of the growing awareness and demand voiced within academia and Arctic communities to rethink scholarly engagement in the Arctic. It is concerned with how Arctic research and researchers can come to matter in new ways by bringing them- selves, their methods, and their engagement to the field of investigation in new and meaningful ways. To initiate a reflexive as well as practical endeavor into what that might mean and entail for Danish research institutions and their researchers, this volume presents some of the most recent experiences with collaborative research in Greenland made by researchers at Aalborg University in Denmark.

During the 1990s Bent Flyvbjerg developed his seminal ideas at Aalborg University on "making social science matter" through the application of phronesis. In his work, Flyvbjerg criticized social science for emulating the natural sciences in its urge to develop objective and reproduceable knowledge. In contrast, Flyvbjerg claimed that the contribution of social science was *not* its ability to produce universal knowledge, but rather to take a practice-based or, as he termed it, phronetic approach to the world and knowledge production (Flyvbjerg, 2001: 61)::

What should be expected, however, is that phronetic social scientists will indeed attempt to develop their answers, however incomplete, to the questions. Such answers would be input to ongoing dialogue about the problems, possibilities, and risks we face, and about how things may be done differently.

Aristotle was the first to formulate phronesis as one of three intellectual virtues important for society. Where *episteme* refers to scientific knowledge that is context-independent and universally applicable, *techne* refers to crafts and differs from *phronesis* in its orientation toward production and is based on practical instrumental rationality. Between these two extremes is phronesis, which is also understood as practical wisdom manifested in the capacity for the correct judgement of action in particular circumstances. According to Flyvbjerg (2001), phronesis "is that intellectual activity most relevant to praxis. It focuses on what is variable, on that which cannot be encapsulated by universal rules, on specific cases" (p. 57).

The exertion of phronesis is based on practical value-rationality, and it therefore revolves around values and how decisions and actions are based on such values (Flyvbjerg, 2004). Accordingly, thinking of the social sciences as a tool to produce value-free knowledge is a misguided objective. Phronetic research aims at action and skill development, connecting different kinds of knowledge, the objective being to provide "input to the ongoing social dialogue about the problems and risks we face and how things may be done differently" (Flyvbjerg, 2001: 61). Through engagement, perceived scientific weaknesses about social science, such as the difficulty in building cumulative and predictive theory, are turned into strengths, as is the local, practice-based, and situated knowledge.

The contribution of social science is *not* explanatory or predictive models, as one would expect from natural science, but understanding. Flyvbjerg argues that we must focus on the problems that matter to local, national, and global communities. Importantly, we must publicly communicate, deliberate on, or find some other way to put the results emerging from such research "at risk:"

> We may transform social science to an activity done in public for the public, sometimes to intervene, sometimes to generate new perspectives, and always to serve as eyes and ears in our ongoing efforts at understanding the present and deliberating about the future.
>
> (Flyvbjerg, 2001: 166)

So how might phronesis add to making Arctic research matter in new ways? And what might this research look like? While acknowledging that we, Danish researchers who are drawing on a Western academic system of thought, are in no way disentangled from power relations, we suggest that a more phronetic approach in Arctic research can serve to produce more practical and situated knowledge. It might also help to balance the power of the instrumental and scientific rationality. In the following section, we describe how the scholarship at Aalborg University and within the AAU Arctic research platform seeks to navigate these waters.

A sensibility for collaboration: Aalborg University, Aalborg, and the Arctic

A phronetic approach offers valuable insights into Arctic research in directing attention toward the aforementioned environmental and political changes for local and global communities and stakeholders, as well as the changing expectations of the role of science in society. But there are many ways to do so. This volume introduces collaborative research as a way of crafting "Arctic research that matters." We do so by offering the experience from a range of Arctic researchers at the Aalborg University research platform for Arctic research, AAU Arctic. As discussed below, it is no coincidence that collaborative, phronetic Arctic research thrives at Aalborg University.

Since being founded in 1974, Aalborg University in Denmark has had strong traditions for collaborating with various actors, including authorities, businesses, NGOs, citizen groups, and communities. With its innovative and pragmatic take on identifying issues at stake and working with problem-solving, the Aalborg model for problem-based learning (PBL) has received global acclaim and a UNESCO chair in PBL. Responsible for coordinating research and collaboration between disciplines and partners in, with, and for the Arctic, AAU Arctic brings this collaborative and problem-based heritage to Arctic research while linking its research activities to its local context.

Aalborg, Denmark's fifth-largest city, has hosted the "Greenland Port" (Grønlandshavnen) for more than a century, meaning that it has long served as a logistics hub for goods to and from Greenland. In particular, the relationship between Aalborg and its twin city, Greenland's capital of Nuuk, is long and important. Similarly, the Arctic research at Aalborg University is marked by a major focus on Greenland. The following chapters exemplify this tradition and the aim of Aalborg researchers to continue and support partnerships and to build mutual trust.

As shipping routes are redirected and colonial relations replaced by new political, economic, and emotional ties, research must also raise questions regarding its own value. Importantly, research should not only emerge from a genuine interest in understanding and engaging with the realities and values of the Arctic, but, to paraphrase Flyvbjerg (2001), should also be responsive to actors and communities to intervene, create new understandings, and deliberate about the future. These shifting configurations and new demands pertaining to research also underpin the work and discussions in the contributions to this volume.

While collaboration is surely relevant to the contributing authors to this volume, *how* it is so assumes many different forms, as does *why* it is so. Similar to the AAU Arctic research platform, this compilation of research accounts stems from researchers representing a broad variety of academic fields, ranging from law and medicine through business studies and planning to cultural studies, ethnology, and anthropology. The contributors draw on their experience with working in Greenland and together with Greenlanders. While few of the contributors are native to the country, most of them have strong ties to Greenland, either through family, research stays, extended periods of residence, fieldwork, or collaboration.

What unites the contributions across the conspicuous differences in topics, fields of research, and even philosophies of science is the commitment of the authors to talking, working, and thinking together with local communities and businesses, with event organizers and participants, with patients, and with students and fellow researchers in the Arctic. While not all of the chapters in this volume draw on social science, the issue of phronesis nevertheless directs their scope and activities. The contributions from the AAU Arctic researchers are characterized and united by their attempts at reflecting on the situated practices of which their research is part and to direct their research toward real and pressing issues and problems. Each pathway to societal impact should start with society, its challenges, and interests—not with research results.

What is in this book?

The book is structured as follows. In the first part after this introduction, different research positions are introduced based on the contributors' personal research accounts. The chapters display very different ways of interpreting and carrying out collaboration in research projects—from ideological change in the role of science in the contemporary Arctic (Reimer) over very concrete and mundane, yet also very impactful interactions (Andersen) to reflexive subject positions (Graugaard). As the chapters show, all of these "collaborative inclinations," whether related to general societal change or shifting concerns of research positionality, each, in a different way, enable the creation of new kinds of knowledge, understanding, and values not warranted by or stemming from "hit-and-run" research (Graugaard, Chapter 3) or, as coined by Henrik Halkier in the foreword, a "fly-in/fly-out" approach.

The ongoing move away from "detachment" in Arctic research is addressed head-on in Chapter 2 in a conversation between Anne Merrild Hansen, head of AAU Arctic, and Minik Rosing, Professor of Geology at the State Natural History Museum, University of Copenhagen, and Chairman of the Ilisimatusarfik Board of Directors. The latter is himself Inuit and an active researcher in Greenland. In their conversation, Hansen and Rosing deliberate on different positions that research can take in relation to society—and also where things get more complicated

More specifically, they also discuss how research can contribute to stimulate a harmonious development of society in Greenland.

In Chapter 3, *Participatory principles in Arctic health research,* Stig Andersen describes his decade-long and ongoing commitment to health in the Arctic. He reflects on public engagement methods, challenges, and experiences applied in the health care sector.

In Chapter 4, Naja Graugaard introduces auto-ethnography in *Arctic auto-ethnography: Unsettling colonial research relations as a way of engaging with a sense of failure* to push for continuous reflection on methodological and analytical choices. In her exploration, Graugaard approaches the concept of collaboration through reflection on what it means to create knowledge. Can it be co-constructed? And if so, how to grasp and act on the implications and limits of such collaborative engagement? Seeing the Arctic research encounter as a space inhabiting multiple relations of power, the chapter discusses the potential of auto-ethnography to inhabit this space of uncertainty and failure. In her account, Graugaard repositions Arctic research from being (seemingly) distant to being "in the midst of things," hereby also recasting knowledge production as a highly situated and power-ridden endeavor.

As illustrated very clearly in these three introductory testimonies, the commitment to collaboration can take many forms. In the second part of the book, we explore this even further by introducing an array of cases based on research projects in Greenland, which in one form or another are characterized by collaborative efforts. In the most general terms, a case study denotes a close investigation of a particular example. The defining feature of a case study is thereby an examination of particular circumstances or the practices constituting a given field of interest. Referring back to Flyvbjerg (2001), we learn that these characteristics make the case study method integral to the practice of phronetic research.

While the merits of the case study method are often evaluated based on their capacity to produce generalizable knowledge, Flyvbjerg argues that there is no script for choosing the right case to secure generalizability. This is actually far from the objective of much case-based research. Cases should not be taken as small, bounded examples of a larger whole. Rather, they should be understood in their own right; as sensitizing devices that may invoke questions about differences and similarities between sites and circumstances under study, as these chapters illustrate. In all their diversity, the cases are united by their sensibility for collaboration and for the interest in what we might term "situated mattering;" that is, to contribute to ongoing Arctic conversations.

In Chapter 5, *Industrial development in Nuuk and Sermersooq: Empowerment through action research,* Allan Næs Gjerding and Ina Drejer introduce the concept of action research in their description of how they used an action research-based approach in a project where the success depended on securing the consent and involvement of actors. They reflect on the methods and outcome, together with the value of local competencies, as well as on how scenario-building in particular served as a platform in their project, for dialogue between stakeholders drawing the attention away from present disputes in favor of addressing common aspirations and related imaginative actions to reach desired futures.

In Chapter 6, *Collaboration to secure relevance and quality in a study of EIA practice in extractive industries in the Arctic*, Sanne Vammen Larsen and Anne Merrild Hansen present their collaboration with various partners from industry, government, and research in a benchmarking study of the regulations related to hydrocarbon exploration in the Arctic.

In Chapter 7, *Critical proximity in Arctic research: Reflections from the Arctic Winter Games 2016*, Carina Ren and Robert C. Thomsen propose and explore different researcher–field relations, discussing their implications for our ability to know about—and with—the Arctic. Their literature review describes how Arctic research in the area of community and identity-building has previously worked in participatory ways and characterizes the potential of a close collaborative approach to research. They propose the concept of "critical proximity" as a valuable research position in making research matter in the Arctic. Examining the learning outcomes drawn from a two-year engagement with the planning and execution of the Arctic Winter Games 2016, they reflect on the positive outcomes, as well as challenges and pitfalls in working with critical proximity.

Collaboration in the universities of today does not limit itself to research alone. This is illustrated clearly in the AAU strategy *Knowledge for the World*, in which the three columns of research, education, and knowledge collaboration underpin and support the everyday operations of the university (AAU, 2015). The third and final section of the book provides two examples of how collaborative approaches are woven into the two other central fields of interaction for Arctic researchers at AAU Arctic: education (Faber) and knowledge collaboration with external partners, exemplified here in an industrial PhD in business (Huppert). These chapters point to how collaboration cannot be reduced to research projects and how it overflows to areas of teaching, learning, and innovation.

In Chapter 8, *Development of jurisprudence research through engagement of students*, Louise Faber describes the work of establishing an education program in Greenland while covering a very personal perspective on the topic of collaboration.

In Chapter 9, *Recruiting and retaining labor in Greenland: A PhD project in close cooperation with local stakeholders,* Verena Gisela Huppert draws on the experience from her participation-based PhD to illustrate how research is locally anchored and activated in close collaboration with local partners in Greenland. She also shows and discusses how participatory research can be conducted toward this end in the Arctic, as well as how roles must be balanced in the course of such collaborations.

In Chapter 10, *Life Mapping: A participatory approach to insights on tourism collaboration in Greenland,* Daniela Chimirri explores how the actual "act" of collaboration is envisioned and mapped within the Greenlandic tourism landscape. Using an innovative technique of life mapping co-created with tourism actors during research, she unravels how collaboration is articulated as an essential element in the daily operation of tourism businesses. While the findings contribute to our understanding of the intricacies of collaborative practices, the technique itself, as argued by Chimirri, also co-creates knowledge and awareness, which ultimately empower local actors to take an active stand in the ongoing discussions concerning tourism development in Greenland.

In our concluding remarks in Chapter 11, we gather the learning drawn from the account in pointing out how further collaborative research can provide a solid base and pave the way for future projects in, for, and with stakeholders in Greenland.

References

AAU. (2015). *Knowledge for the World* (Aalborg University Strategy 2016–2021).

Arctic Monitoring and Assessment Programme. (2017). *Snow, water, ice and permafrost. Summary for policy-makers.* Oslo: Arctic Monitoring and Assessment Programme.

Carson, M., and Peterson, J. (Eds) (2016). *Arctic Council. Arctic resilience report.* Stockholm: Stockholm Environment Institute and Stockholm Resilience Centre.

Deeb-Sossa, N. (Ed.). (2019). *Community-based Participatory Research: Testimonies from Chicana/o studies.* Tucson, AZ: University of Arizona Press. doi:10.2307/j.ctvdjrpp5.

The First Nations Information Governance Centre. (2014). *Ownership, control, access and possession (OCAP^{TM}): The path to First Nations information governance.* Ottawa, ON: The First Nations Information Governance Centre.

Flyvbjerg, B. (2001). *Making Social Science Matter: Why Social Inquiry Fails and How It Can Succeed Again.* Cambridge: Cambridge University Press.

Flyvbjerg, B. (2004). Phronetic planning research: Theoretical and methodological reflections. *Planning Theory & Practice,* 5(3), 283–306.

Hansen, A. M., Tejsner, P., and Egede, P. P. (2016). Traditional knowledge and industrial development: On the potential use of indigenous and local knowledge as a resource to assess competencies in Greenland. In R. Knudsen *and* M. Jacobsen (Eds), *Perspectives on skills* (152–166). Copenhagen: Greenland Perspective.

Iglič, H., Doreian, P., Kronegger, L., et al. (2017). With whom do researchers collaborate and why? *Scientometrics,* 112: 153–174. https://doi.org/10.1007/s11192-017-2386-y.

Keskitalo, C., and Sothcott, C. (2014). Globalization. In G. Fondahl and J. N. Larsen (Eds), *Arctic Human Development Report II: Regional Processes and Global Linkages,* TemaNord: 397–421.

Koivurova, T., and Lesser, P. (2016). *Environmental Impact Assessment in the Arctic: A Guide to Best Practice.* Cheltenham: Elgar.

Lawson, H. A., Caringi, J. C., Pyles, L., Jurkowski, J., and Bozlak, C. T. (2015). *Participatory Action Research* (pocket guide). Oxford: Oxford University Press.

MacKinnon, S. (2018). *Practising Community-based Participatory Research: Stories of Engagement, Empowerment, and Mobilization.* Vancouver, BC: UBC Press.

Ren, C., Van Der Duim, R., and Jóhannesson, G. T. (2017). *Co-creating Tourism Research. Towards Collaborative Ways of Knowing*. Abingdon: Routledge.

Rink, E., Reimer Adler, G., et al. (2019). *Community based Participatory Research in Greenland*. Nuuk: Ilisimatusarfik.

Saxinger, G. and First Nation of Na-cho-Nyak Dun. (2018). Community-based participatory research as a long-term process: Reflections on becoming partners in understanding social dimensions of mining in the Yukon. *The Northern Review*, 47: 187–207.

Stephen, K. (2018). Societal impacts of a rapidly changing Arctic. *Current Climate Change Reports*, 4(3): 223–237.

2 Telling the good story

A conversation with Minik Rosing on research collaboration and research in Greenland

Anne Merrild Hansen

Research collaboration can be seen as being about creating synergies, about learning and educating, informing, warning, persuading, sharing, mobilizing, and adapting to change. The different individual experiences, mental and cultural models, and underlying values and worldviews shape the researchers' approach to their research. A prerequisite for collaboration is communication between partners. A simple transmission model of communication can be seen as comprised of a messenger, who transmits a message, through particular channels, to specific audiences. This describes a one-way process and assumes that partners are passive, simply receiving the information conveyed by the message of the researcher. When we investigate collaborative research methods involving more complex interrelationships with people within our research field, we go far beyond simply informing partners of our findings. But why are researchers in Greenland increasingly conducting research in this manner? What is the role of the researcher in the relationship with other stakeholders? And what is the purpose of collaboration? I have invited Minik Thorleif Rosing, a respected Greenlandic geologist, professor at the University of Copenhagen, and founder of Greenland Perspectives[1], for a conversation to reflect on these questions and share his views on research collaboration in Greenland.

What research and science is all about

ANNE: What motivated you to become involved in research?

MINIK: I never intended to become a researcher when I was young. I've always been very interested in rocks and mountains and landscapes. I found out that geologists spend a lot of time outdoors. That was an important part of my motivation to become a geologist. I've always been interested and fascinated by science. It was this interest that drove me into the field—never the idea of becoming a scientist.

ANNE: You have been engaged in many different types of projects, such as the project focused on the age of the universe about *what was*, and you were also involved in projects about climate change, which is more about *what is* or *what might come to be*. And then you initiated *Greenland*

Perspective, which is looking ahead and trying to develop new initiatives and innovations for the future. How does it all relate?

MINIK: For me, it's all connected. Ultimately, it's all driven by curiosity and passion. I find that science is about unfolding and telling good stories. We all know that science can be applied for the benefit of society; it has the potential to create change. The awareness that what you know can make a difference makes you want to contribute and make something that has a positive influence. To contribute to… if not a better, then at least a society, which is not worse than the one we already have. Climate is a central theme in this respect, and climate is closely related to the story about the Earth and how the earth systems function. In geology, we can study millions of years of history and find out what goes wrong. When we look at the findings, then the biological history uncovers stories about species that succeeded in being successful in a given environment at a certain time. But the curse is that if they become too successful, then they manage to change the environment they are in. And then it ends in disaster.

ANNE: Do you think this pattern is repeating itself again now?

MINIK: Yes. But we have a straw to hang on to—and we know about all the other times it has happened, and we're capable of doing something different to change our path. This is a unique geological event in history, where there is a species heading toward catastrophe, which is capable of changing this development if we want to. I find this very fascinating.

ANNE: These are huge issues that you investigate and communicate: about the universe and climate change. I suppose that is somehow in contrast to what you do regarding the glacier rock flour, which is a very practical study focused on innovation?

MINIK: For me, there's a good connection. Greenland Perspective grew out of the research I was involved with together with many other researchers. But my experience, and this is also the experience of others I learned, is that it's really hard to make the research relevant for Greenland—for the society, to benefit them. If the purpose of doing research is to elaborate on the stories we extract about the world and to learn to live in a more sustainable manner in our surrounding environment, then you can become frustrated when you're only working with details in the academic literature, which is then maybe read and cited by 20 other researchers. It may get more attention, but it becomes frustrating not to be able to communicate it directly to someone who will benefit from it. On the other hand, scientific literature always focuses on arguing for why data and results are legitimate—but it never tells you why we do it or what it is that we don't know. You have to do something different to reach that goal. Another thing is that there has been dialogue in recent years about where Greenland should go in the future, and the geology of Greenland has been at the center of this discussion as a potential mining country. I found that the knowledge about geology was insufficient if it was not coupled with knowledge about the society, including different disciplines such as law, economics etc. So you could say that the "For the benefit of Greenland" working group was a pilot study, trying to not only make science but to gather knowledge and couple it to make sense to the people who had to make decisions in this regard—about the future of society.

ANNE: So it's about going from being a producer of data to also becoming an interpreter and disseminator of knowledge?

MINIK: Yes. Exactly. It's about qualifying and bridging the jump from information to knowledge. The problem with most scientific reports is that they cover such a limited area of the world. For normal people it can be very hard to see what they can to do with this knowledge in the broader perspective. Therefore, you need to bring lots of different sciences and knowledge together and boil it down together to an extract of what it really means. It takes a very long time and It might not be in a very high scientific standing, but it's very satisfying because you finally manage to activate all that you have gathered over many years and make it relevant to people. I think many appreciate it and welcome it. It isn't necessarily the decision-makers—but people from society in general in Greenland and Denmark give a lot of positive feedback.

The role of an engaging researcher

ANNE: Do you think you would have reached the same point and assumed the same role if you didn't have a history in Greenland?

MINIK: Yes and no... I find it to have been determining for my role in two ways: First, in relation to legitimacy—I'm allowed to have an opinion because I'm considered close but still at a certain distance. And secondly, even if I don't live here on a permanent basis, I still have an understanding of the local context. People coming to Greenland from other countries or parts of the Kingdom of Denmark can also gain a good understanding of the local context, but those coming from far away will find it more difficult to figure out which topics are relevant and how they can be boiled down to become useful.

ANNE: Many geologists have passed through Greenland over time. But the desire or interest that drives you to use your knowledge of the geology to do something that can make a difference for Greenland—does that have to do with your personal background?

MINIK: Yes, I think so. Many who work here, not for lack of will—they just don't think about it. It doesn't occur to them. For them, Greenland is a research object. And that's also OK. I see no problem with that attitude. But for me, it has become part of a criterion for success to take the knowledge we have about Greenland and translate it into something that creates a broader enthusiasm in society. I actually think that this is the purpose of all of the stories shared according to the cultural tradition of Greenland. It is the Greenland way of constructing a story—it not only has to do with knowledge transfer, it's about creating a mood and sharing joy.

ANNE: So as I hear you, the relation between you and society has both helped you to identify what is relevant to investigate as well as created a drive. I also recognize this from myself: that need to make a difference through what you're doing. And for it to contribute to something more—long-term solutions, rather than simply data production. Something that I personally noticed—which I find a challenge connected to the close relationship I have with people in Greenland—is when people expect me to translate science into personal recommendations. For example, when I was in Narsaq as part of an expert team providing information about uranium mining. After we made a presentation at a public meeting about the potential impacts related to a proposed mine in the area involving rare earth metals and uranium mining, we had a Q&A session and a woman stood up and asked, "Would you yourselves live in Narsaq if the proposed mine becomes a reality"? The other scientists replied that they couldn't answer, since it was more a personal or even political question than a scientific question. They were able to uphold that distance. But I felt I had to relate to the question. I grew up in Narsaq, and I knew the woman who had posed it. I felt that it would be dishonest in a way if I didn't dare to have an opinion.

MINIK: I think it's extremely important as a researcher to dare to have opinions. I don't appreciate the idea that as a scientist you need to be a clinician. It's OK to have opinions. You just have to be very aware when you're answering a scientific question and it's a personal question or opinion. Those are two very different things. In my opinion, it would be appropriate for you to say whether or not you would live there—but let them know that this isn't based on your scientific insight or guidance; it's a personal opinion.

ANNE: I agree. And that was actually what I did. My colleagues had just underlined in their presentations that the mine wouldn't pose any particular danger to the people living in Narsaq. And even though I understood the science behind the presentations and believed their conclusions, then I still think that if such a mine as the one proposed is developed in Narsaq, even while my father still lives there, then it would no longer be the same town I knew from childhood, and it probably wouldn't be a place where I'd want to live.

MINIK: There can be all sorts of parameters, which aren't scientifically quantifiable. And you're of course allowed to be influenced by them as a person. As a researcher, you must not be afraid to have and express an opinion as long as you are clear on when it's a matter of personal opinion.

ANNE: Can being a Greenlander yourself make it more complicated to be a researcher? Can it create challenges?

MINIK: Maybe... Actually, I don't think it's a problem. Many Greenlanders approach me when I share my opinions and thoughts and express their appreciation. Some say: "being a Greenlander and researcher at the same time allows you to have a qualified opinion." But my independence of the system in Greenland also plays a central role in this regard. If you're completely integrated in the greater social network in Greenland, then you might actually have a more limited degree of freedom to express a critical view than otherwise.

ANNE: I also find that it can be beneficial to have the relation where you know the society in Greenland and are considered a part of it but you're not completely woven into it.

MINIK: I find it interesting that it's something that people reflect on and appreciate—the fact that it's possible for me to be a part of the society, yet to be somewhat independent.

ANNE: But then what about the people who come from outside of Greenland—like my colleagues from Aalborg University who want to go to Greenland to conduct studies? Do you think it's a prerequisite that they should first develop a relationship to the Greenlandic society? To be able to form a qualified opinion?

MINIK: I think it depends on what you're researching and what you're doing. Your approach should be reflected on the product label of the results afterwards.

ANNE: So it's important to be transparent about how you conduct your research and what it means for your results? If you apply a clinical view as an outsider, this should be made clear. If you're from Greenland and this affects your research, that should also be clear. And anything in-between.

MINIK: Yes. Then we all know what we're dealing with. We all have blind spots. Even insiders can have trouble seeing the forest for all the trees. I mean, it can sometimes be hard to see something if you're a part of it. In European history, some interesting events contributed to the under-standing of what Europe looked like during the Renaissance. For example, documents from a Japanese sailor who got lost and travelled through Europe to get back home. He kept a diary where he wrote about all the things he observed during his travels. This is good material about what society looked like during this period in Europe. The European writers saw their cities from the inside and therefore didn't mention the many issues brought up by the Japanese sailor. We also know this ourselves when we work in other parts of the world or if people from outside come here to work with us—they sometimes question how you do or see things, and that's really healthy for all of us; to sometimes be challenged in relation to how you should or could do things differently. Sometimes we think that things are a certain way, but there are so many different pos-sibilities. It's a question of casting light on a society. When you do research, it isn't about saying something about how a society should develop or what they should do; rather, it's about providing information about the different paths available. I don't think it's our role to say what the right or wrong way to do things is.

ANNE: I think that it's also about pointing out the consequences of implementing different options—if we can see that certain decisions will push in a certain direction.

MINIK: I think that's really important. If you're a geologist, for example—if you find a place with a lot of gold, you can't start speculating about all the things Greenland could do with this gold. You should say that you've found this amount of gold in this location, and then leave it to others to decide how to use that knowledge. That's the balance you need to keep, also when talking to the media about something. For example, when you talk about climate change and you're saying something about how much inland ice is melting, then you should avoid speculating about what that might lead to and leave it to others. If you're the one giving the numbers, then you shouldn't also present the implications—in my opinion—in these relations. You have to choose who you are.

ANNE: I see your point. But isn't there a conflict with what you said about how you shouldn't only be a data-producer but also share your knowledge and insights?

MINIK: You can share your data without providing instructions about what precautions to take.

ANNE: So for researchers, it's about saying something about what you *can* do rather than what you *should* do?

MINIK: You could say: "this amount of ice is melting from Greenland. That's three times the amount compared to anytime before, and that will mean this and that." But you shouldn't say: "this is a catastrophe and here is how you should act." There are a lot of grey areas. But in general, you shouldn't be afraid of getting close to the border. Sometimes it's better to get a little too close rather than creating a distance to the world.

Shaping the future

ANNE: Can and should you as a researcher shape the future? And how do you do so?

MINIK: Yes, you should. Even with a lot of trouble. No. Actually, I don't think that all researchers should have that as their goal. We're all different as researchers, and this is a prerequisite for having communities/ societies that we all have something to contribute with in different ways. Research communities also need some nerds and some who are only interested in making technical gizmos. That's how it should be. But that said, I still think you have an obligation if you have insight of meaning to society—then to pass it on to someone who can use it. I really think that's important. You can't sit back with important information and withhold it from others. I think the desire to change something to the better for society can drive you. But it isn't up to you to determine the direction. You should explain the nature of the parameters upon which a decision is based.

ANNE: It sounds like a fine balancing act?

MINIK: Yes, it is. Indeed, it's a very difficult balancing act. We all know it. It's a bit of the same situation when one of your friends is having a problem and you want to help. You can talk to them and listen and understand—but you can't provide the solutions. It's about helping them reflect on the implications and maybe see their situation from new perspectives to be able to make a decision.

ANNE: So—if we're thinking this in relation to our reflections on how to interact when doing research, does it then mean that there isn't any "one-size fits all" model? If you're saying that all researchers are different, and they should be allowed to conduct research in different ways?

MINIK: Definitely. There are people who would never speak in public, because they can't. I think it's extremely difficult to make guidelines and rules for how researchers should act. Some have the skills, sense, capability, and desire to do something—and others to do something else. Researchers undoubtedly perform better if they, within the limits of reasonability, get to do research the way they want to.

ANNE: There are different scientific disciplines. Is it equally relevant for all disciplines to take the responsibility to engage with local communities?

MINIK: I think it is equally relevant to consider for all disciplines. Then it's up to the individual researcher to think about if and how it's relevant for a particular project. If you do anything that is connected to the outside world, then you have an obligation to make an effort to share your insights.

ANNE: Is providing information to society the primary function of research?

MINIK: The primary role of doing research is to satisfy our curiosity. That's a central part of being human. I see it as if the research has no specific purpose. Just like opera has no purpose or art—but we would rather not live without any of them. Research has the side-effect of making you realize which resources and opportunities and limits there are to our world. The other disciplines do the same. They also contribute to creating an impression of our surroundings, which helps us live in them. There is no fundamental difference between disciplines.

ANNE: So it's all saying something about ourselves. It's about identity and creativity and identification of the world and how we interact with it—and the implications this has for the future?

MINIK: Research is also a social activity. And it's just as important as other social activities.

ANNE: So you're telling me that we should conduct research because we think it's awesome? And then when we occasionally stumble over something that may be important for others—then we should remember to pass it on to those who can make some informed decisions?

MINIK: Yes!

Science diplomacy

ANNE: Does science diplomacy play a role in Greenland?

MINIK: Yes, it does. Exactly because of what we were just talking about: research being a social activity between people who share some joy, curiosity, and other things. And as the world has so many parameters, the odds of finding someone in your nearby community is relatively small. And therefore you find international networks of people that you get to know really well—and then you get new insight into how other countries work and think. In that manner, it creates a cross-national and cross-cultural understanding. You get an understanding about what happens elsewhere and if something is building up. It tells you a lot about what's going on in the world. Another thing is that research often develops its own language in the individual research communities. It means that we, by engaging in these communities, become capable of hearing, reading, and understanding the knowledge produced by others. Other parts of society may not be able to readily consume this knowledge, so we have to translate this knowledge into something useful for our society. Again, this is about being able to contextualize what's going on in relation to our own society. And back to the question of whether your connection to Greenland means something for research dissemination. And it *does* mean something in relation to being able to understand others' data—and then making it available in an understandable manner in your own context.

ANNE: Maybe it's also about seeing patterns identified in others' research that are also present in your own community. When you say that data and knowledge is an important communication platform, do you then also believe those who say that research is contributing to upholding peace in the Arctic?

MINIK: For sure. I'm convinced that research has contributed to the Arctic—until now—being a united phenomenon. The Arctic Council is built up around joint environmental programs. So you could say that some of the international fora that are regulating activities in the Arctic have developed from research activities. This itself is research diplomacy. Politics is behavior regulation. And regulating behavior based on knowledge is always better than if it is not. Therefore, I think that when researchers who are informing politicians in one country are friends with researchers informing politicians in another country, then there is a chance of building up a shared understanding of reality. Alternative realities can lead to trouble.

Indigenous knowledge

ANNE: Yes. It's good when we agree on which reality we're in. Then we can push in the same direction when we want to solve a common problem. Is there anything that is particularly important to consider when conducting research in the Arctic?

MINIK: You need to wear warm clothes [laughs]. I think there is something that is unique to some degree—but which is also the situation in a few other places: that the indigenous peoples constitute the majority of the population. This means that there are some concerns you need to take— and others that you should avoid taking. On the one hand, you have to acknowledge that this is the case, but you also have to be careful not to become trapped in the perception that everything is so special and different that it makes it different from what happens in other parts of the world, both in relation to demographic development and other issues. You should be aware of not concluding that when something happens to you—if you're hit by a certain wave or trend, that it's unique for you due to your location. The exact same trends are sometimes present in other parts of the world. Of course you need to respect the views of Indigenous Peoples and to grant consideration to them. But you shouldn't be so considerate that you avoid addressing and sharing important messages.

ANNE: Can you give me an example? What should you do to be considerate and what should you avoid?

MINIK: The only thing I think researchers should do to be considerate is to take people seriously—who they are. Listen to what they're saying. But also make it clear if you disagree with their views. It becomes a problem if it's considered unacceptable to openly disagree with Indigenous Peoples—if a researcher doesn't express their views in the dialogue with the people they're talking to but then just shake their heads when they get back home. Sometimes we also use different definitions of concepts, which can be problematic. Take indigenous and traditional knowledge, for example. There's so much of it, as I see it. And yet there's so much disagreement about what it is exactly and what to do with it. My thought is that if knowledge exists, then it should be used. It must be possible to subject it to the same tests as all other forms of knowledge. It might relate to something other than what you might expect. It often relates to topics such as population biology. Like in relation to the number of nar- whals in an area. A local might say that there are a lot of them—and there might be in that particular area. But if the whales are gone in the rest of the world, then it becomes problematic to rely on this knowledge alone. Then it also becomes a question of the local versus the global. But you shouldn't question how many there might be where that person observed them—or question that knowledge. You should recognize it as a single data-point in a much larger study—not that there is disagreement regarding the narwhale population. There isn't.

ANNE: So—it's about upholding respect in the dialogue? And researchers should dare to raise discussions?

MINIK: Yes. You should talk to people with honest interest and find out what they think and say. And then you must let them know what you think and say. And if there is disagreement, then you have to try to locate/

identify what the disagreement is rooted in. You might be talking about two different things. Sometimes, even if there seems to be disagreement, both parties might actually be right. We all have—and I find this a bit amusing—a tendency to think that if there's disagreement, then there is one who is right and one who is wrong. But if two people disagree, then they're often both wrong. If two people meet and agree on something, then there's at least a small chance that they're right about it.

ANNE: My impression is that the people I've met who are living off the land and sea in Greenland are highly dependent on interpreting the signs of the surrounding environment to insure a continuation of their way of life. And that they possess in-depth knowledge about environmental variations and seasonal changes—and how these factors influence the harvest, as you also mention. But I think that there's also knowledge that relates to the culture and how to uphold the quality of life in an Arctic environment. For example, I've heard about researchers who come to observe and ultimately conclude that children are neglected because they stay outside after 10pm on normal weekdays. But when I was a child in southern Greenland, we all ran around outside after 10pm in the summer evenings when it was still light!

MINIK: Yes, yes, yes! I think this is one of the paradisiacal things in Greenland—the sound of children playing in the night under the midnight sun. It's the best ever! It relates to the perception of what's considered "children's rights". If children want to play in the sun, who are you to say that they cannot?

ANNE: The criteria for what is considered a good life for a child in Greenland may vary from place to place. And you can measure and quantify this differently in different places. I think that we as researchers somehow have an obligation to try to understand the reasons behind behavior—and what can be concluded from different actions. That's also local knowledge. If you come from the outside, you may be able to point out issues and describe the reality as you see it. You may see certain things that the locals do not see. But you might also point out indicators and interpret them and reach conclusions that are considered wrong from the local point of view.

MINIK: That's why you need dialogue—where you can talk openly about what it is that we see. You might conclude that it's super cool that things are a certain way. But with the ongoing developments in the world, that might not continue to be the best solution. Maybe you're living in a town where you have to get up every morning and go to school—it might not be healthy. Maybe the late-night playing should be kept for the summer holidays. We only become smarter by engaging in these types of dialogues.

ANNE: There may also be different views and relevant solutions in different parts of Greenland. It may not be as safe for a child in the capital, Nuuk, to go out at night where there are many cars and people who are strangers to the children. Whereas it may still be relevant and possible in smaller settlements.

MINIK: I don't think that safety has ever been the big issue in the upbringing of children in Greenland. Greenlanders don't necessarily equate life-length and life-quality. We think that it's important that children have the freedom to do all sorts of things. And then sometimes it goes wrong. Rather than keeping them locked up. There's a good reference to this phenomenon in Knud Rasmussen's collections of essays, which includes a story about a man who lost his two oldest sons but wasn't afraid of losing his third son. When asked about it, the father replied, "you can't live in fear."

ANNE: On the other hand, you might also say that those who then survive the free life—or manage to succeed in the dangerous games—they might actually have a greater chance of living a long life in the end because they learn how to act in the environment.

A good story involves collaboration

ANNE: We have now talked about researchers having an obligation in relation to passing on knowledge and the importance of researchers trying to understand and engage in dialogue together with the locals about the context, research, and findings. But other than what we should do for the sake of the people, is there then also something to gain for the researchers when they engage more with locals?

MINIK: There is always something to gain by conducting research with open eyes. Engaging opens up the perspective. If you don't engage and listen to what the locals know and have to contribute, then you lose a lot. It might not lead to a different result in the end, but just the joy of meeting and sharing with the locals is a benefit. I think we're stupid if we don't do it—it shouldn't necessarily be an obligation, but it's always good to reflect on what a project could gain from the engagement to make it meaningful for all parties.

ANNE: So indigenous knowledge doesn't necessarily have to be interpreted as something that can be included in a report or scientific article. It can also take place verbally in the dialogue between researcher and locals. And potentially be brought into action that way.

MINIK: It's like when you're making stew. The potatoes make up most of the substance, but most of the fun is in the spices—which are a very small part of the whole. Engaging [with the local community, ed.] might not change the research substantially, but it adds nuances, perspectives, and content to the research process that makes it more fun and interesting. Things are also more interesting to read if the local perceptions are integrated. And the better an article you write, the more readers you reach.

ANNE: Then we end up where we started—with the importance of telling a good story.

MINIK: Yes. What really matters is the good story. The researcher's job is to tell the good story. And the best story is developed together and shared with a broad audience.

Note

1 Greenland Perspective is an multi-disciplinary, multi-national research initiative, which works on the basis of a set of precepts aimed at the activation of research projects into society: https://greenlandperspective.ku.dk

3 Participatory principles and privileges in Arctic health research

Stig Andersen

Introduction

Completing that transatlantic flight and getting off the plane in Kanger-lussuaq was a mind-blowing experience. Stepping out into an Arctic afternoon in early October, my lungs feasting on the oxygen-dense, dry air at −19°C, a clear blue sky, and the sun low on the horizon giving a faint burning feeling on my face. My first experience of Greenland. My first time on the brim of the Ice Cap. My first view of the Arctic light. Different. It is fair to ask about that light if you have not seen it. For it is special, and this chapter is dedicated to illustrating that point and some of the advantages and limitations that come with Arctic residence, resilience, and reference.

The data to support my claims pertaining to the Greenland environment during the Arctic autumn and winter stands on three legs: the cold, the snow and ice, and the light passing through the atmosphere. The Arctic winter temperatures are low, which reduces the capacity of the air to hold water (Figure 3.1). You feel this in the dry air that you breathe. The lower vapor also reduces the absorption of light passing through the atmosphere, which explains the high-intensity light reaching the surface of the Earth at the Poles. The snow and ice reflect this light, resulting in intense and immense light during the few hours of daylight in Greenland. Hence, Inuit produced early sunglasses from whatever was available to prevent snow blindness (Andersen et al., 2013a), and cataracts remain quite common to this day among the hunters exposed to natural light in Greenland. Thus, the visual impression that hit me on that early afternoon in October tricked my brain. Black stood out in different shades and added an intriguing contrast to the red glow of the snow—a sight that I had never seen the likes of before and which has since encouraged me to consider various aspects of life in the Arctic.

Inspiring a medical doctor in Greenland

Awareness of the fact that man interacts with and must adapt to the environment triggered reflection on the influence of the frigid Arctic environment

Figure 3.1 The vapor condenses immediately above the sea when the ambient temperature goes below -18°C to -20°C. This is particularly easy to observe when the sun is low.

during my initial period of employment in Greenland. Working at Queen Ingrid's Hospital (the referral hospital in Nuuk) provided an opportunity to observe disease patterns (Andersen, 1999) and stages of the disease, together with a glimpse into the relation to ways of living. This has developed over the years, providing insight into the relation between differences in culture and how to handle and cope with disease, together with an appreciation of my limited understanding of these differences and their implications. This has run parallel to my gradual understanding of the society and the people—or at least an awareness of my limited insight. A pivotal element has been colleagues and other health care professionals of Greenlandic origins. This expanding network has allowed for discussions of contrasting conditions that have sparked an understanding of the differences in the setting for life; and, hence, disease.

Curiosity drives learning and research, and unsettled health issues are legion in everyday clinical practice. Thus, some pondering on clinical observations led to an initial paper based on a case series describing an observation of frequency of disease that was puzzling (Andersen et al., 2013a). A skilled colleague of Greenlandic origin was of invaluable support in the subsequent expansion to a full-scale, cross-sectional, population-based survey to provide knowledge on the topic (Andersen, 2005). This survey has provided the basis for PhD projects and a series of reports based on further discussions in everyday clinical practice.

The true value of this colleague's support may be tacit, but nevertheless illustrates the importance of such interaction. Collaboration and inspiration support the optimization of the benefits from the data and, hence, the effort put into the work by the investigational team, the participants, and the funding bodies alike. The benefits are diverse, some of them obvious.

Some of the less obvious, perhaps even tacit, are benefits related to the social norms in written and verbal responses, guidance with respect to recruitment, and details regarding the setup and design of the questionnaire. The early involvement in the phase prior to initiating data collection supported a high participation rate. These are rarely above 50% in such population-based surveys, but ended up at 95% in the initial population-based survey, covering 1% of the total population of Greenland (Andersen, 2005). This unique survey lends appreciation to cultural understanding and linguistic insight beyond plain translation. This would not have happened without the participation of local resources and insight that provide to be a crucial support to the reliability of the study to portray the population surveyed.

A frame for participatory principles in medical research in Greenland

The noticeable success achieved by engaging local resources in the first systematic clinical health survey conducted in West and East Greenland set a framework for future studies in Greenland. The recruitment of local resources emerged from an awareness of and curiosity regarding real-life clinical problems that were easy to relate to with insight from everyday clinical practice.

The next initiative came from a period of employment in North Greenland. The marked seasonal changes—not least the shift from 24-hour darkness to the midnight sun—raised questions as to the influence on human circadian rhythms. The study drew inspiration from how some of the knowledge leaned on in this regard includes how working night shifts is associated with disrupted physiology and mortality. Thus, engagement with local staff inspired the set-up of a study of seasonal change among inhabitants 400 kilometres north of the Arctic Circle (Andersen et al., 2013b). The repeated sampling in a distinct group of people made it possible to add new perspectives and aspects along the way. This investigation thus developed from a study of pituitary response to also include calcium metabolism, bone density, and some aspects of ageing in a rural Arctic society. The latter study depended on nursing home staff and has evolved into a nationwide study that is currently in progress. The former has spurred a series of surveys conducted by young local colleagues.

Recurring periods of employment at Greenlandic hospitals has led to my working with a number of colleagues who saw an interest and found the time needed for spin-off studies. Hence, differences in calcium metabolism led to thoughts on bone-specific diseases. These led to a questionnaire survey on disease risk factors conducted by a young colleague (Jakobsen, 2013) and to radiological surveys conducted locally with limited means (Sørensen, 2015), but with results supporting an Arctic approach to disease management (Fleischer, 2017). These health surveys should therefore have an impact on disease management.

Medical research resources and the dissemination of results

The implementation of research results often poses a major challenge. There is a continuous risk that research ultimately has a limited impact, owing to an inadequate focus on disseminating the results. Such results might end up in a binder or a dull-looking book on a dusty shelf. This would not be appropriate with the limited body of health research conducted in Greenland and with the efforts made by the participants in a limited-sized population.

An important additional benefit emerging from engaging local resources has been the enhanced access to resources that proved valuable in disseminating and implementing the results. The interest among the media and staff was enhanced when results and perspectives were delivered locally, in Greenlandic, and by the potential users of the results. This further strengthens and underscores the importance of conducting research together with local actors and the support of local resources.

A fixed focus on finding ways to disseminate results requires openness to media and, at times, an active approach. The research results have therefore been made available to the media in ways that should inform the not-so-knowledgeable reader, listener, or spectator. This process adds to the understanding of the involved staff of the potential benefits of the results and ultimately therefore also to their implementation.

Research in Greenland is often also of interest to the world more generally, and the presentation of results at international meetings is important. The medical staff in Greenland must adapt to both Greenlandic and Danish, and engaging in discussions in English would be their third language. However, this has not been an issue with the Greenlandic medical research talents with whom I have had the pleasure of working, and senior researchers could thus disseminate their research to a broader target group, encompassing locals and an international audience alike by presenting results in English.

Setting the scene for the continuous development of Arctic health research

It was time to mark the event after ten years of research in Greenland with a formal opening of the Arctic Health Research Centre in Aalborg. The official opening was held in the Aalborg harbor. Aalborg Harbor and Royal Arctic Line hosted the event, which attracted some measure of attention in Greenland. This led to a visit by the Health Committee, which represented the Government of Greenland. This support was highly valued and in itself encouraged further commitment to Arctic health research by the research group in Greenland.

Figure 3.2 Presentations by Greenlandic medical doctors at international conferences have been successful.

In the following year, the newly established Arctic Health Research Centre served as the setting for a symposium held in North Greenland. This symposium gathered research experts from seven countries from around the globe and became a forum for the discussion of ideas and hypotheses on the potential influence of the environment on mankind.

The interest in the Arctic Health Research Centre in Aalborg led to contact with Greenlandic medical students and the organization for Greenlandic medical students in Denmark. This connection further led to Greenlandic medical students writing their master's degree theses on topics of direct relevance to Arctic health (Rex 2012; Noahsen 2013), to a research sabbatical spent in the Arctic (Jakobsen, 2013) and to PhD studies in Greenland on Arctic-related health issues.

PhD studies in health and the Greenland academic environment

One medical student demonstrated an exceptionally high degree of dedication, which led to a PhD study that extended the work he had carried out for his medical degree. It was based on data from the initial population-based study described above (Krarup, 2008), together with long-term follow-up on individual participants. He also conducted a parallel study

Figure 3.3 International and Greenlandic health researchers and clinicians partici-
pated in a symposium held in northern Greenland to discuss the impact of
the environment on mankind.

among Greenlanders in Denmark and contributed to the monitoring of
disease occurrence in a clinical study at ten years subsequent to the initial
study (Rex, 2012).

Another young Greenlandic doctor is working on a PhD study focusing
on the influence of the Arctic environment on metabolism. The path and
construction are similar, as it started with an excellent master's degree thesis,
demonstrating a talent for research leading to the PhD (Noahsen, 2013).
Thus, the early support for research allows the identification of research
talents. This is similar to other areas but complicated by studies abroad, as
the students are encouraged to conduct their research in Greenland. Hence,
a dedicated focus on spotting and supporting research talent is required.
This is supported by keeping these students in groups that are in contact
with relevant research environments.

An additional PhD study investigated the relation between inflammation
and diet, persistent organic pollutants, and vitamin D (Schaebel 2013,
2015, 2017), which extended the hypothesis of inflammation and atherosclerosis that has gained some attention over the years. This contributes
knowledge and access for Greenlandic medical students in an academic
environment.

Figure 3.4 The happy smile comes from the success of a long day's work with physical examinations, questionnaires, and urine and blood sampling from participants in a cross-sectional follow-up study.

The aspiration has been to support and consolidate an academic environment to support health care research in Greenland. The academic environment is of limited volume, and every contribution is valuable. Moreover, the opportunities available to medical doctors are not particularly diverse, and constructing positions combining different academic institutions in Greenland might be necessary. Figure 3.5 illustrates a collaboration between the referral hospital in Nuuk and the universities in Greenland and Aalborg, Denmark. This collaboration provides a basis for specialization in general medicine alongside a PhD. This collaboration supports an academic environment at the University of Greenland in an exchange that feeds academic thinking into the clinical environment that is beneficial for both institutions.

Upcoming studies of the unusual

An inspiring academic environment is open to and allows an unsual idea to develop with guidance, even if it may not initially appear particularly outstanding. Thus, there should always be room and an eye for the unexpected finding and an alternative view.

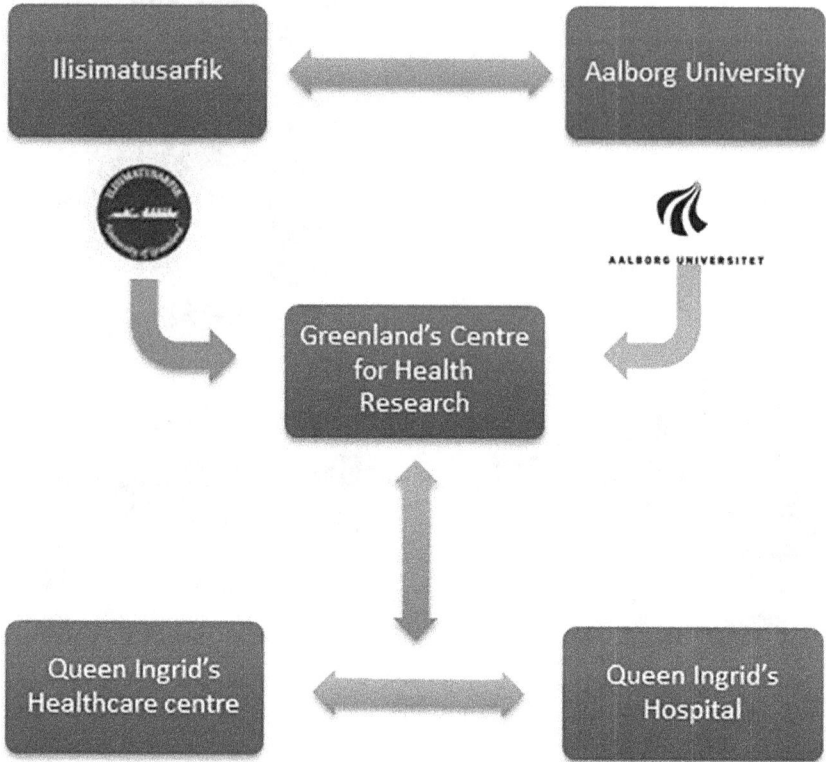

Figure 3.5 The collaboration between the University of Greenland, Ilisimatusarfik, and Aalborg University supports the PhD study for a joint degree, while the hospital and health care center in Nuuk provides opportunity for simultaneous specialization. The collaboration is facilitated by Greenland's Centre for Health Research.

Such processes might begin with a coincidence, the odd observation, a clinical mystery or question, or the keen interest of a member of the academic group. In my case, the latter two occurred jointly and are presently in progress. The starting point was a clinical question concerning dementia in the ageing Arctic population, which was triggered by a study at nursing homes, which found resonance in the clinical setting with employment at the referral hospital in Nuuk. The opportunity came with a Greenlandic medical student who was able and willing to engage in the work for her master's degree thesis. The first step addressed the use of tools developed and validated in non-Arctic settings. This work has progressed into a keen interest in developing a scheme and tools to accommodate the challenges of the vast geography of Greenland and the cultural aspects that are of importance in deciding on the cognitive performance of an individual from Greenland.

Yet another medical doctor is therefore following this path (Noahsen, 2019), which sets the scene for participatory research that promises success for the research topic at hand, the health care staff, the health care system, and for the population and society of Greenland more generally. This could be a path worth following for other Arctic societies.

Conclusion

Being a medical doctor in Greenland holds challenges, including the differences, strengths, fragility, and the complexity of Arctic life. I succumbed, and it became a path to "lifelong" engagement in the support of the development of both health care and academia in Greenland. I believe in participation and collaboration. The experience of working together with and involving local resources in the work is essential, as it strengthens the researcher–field relationship. The importance and relevance of engaging with and developing the available resources in Greenland have been crucial and are of the utmost importance for research, implementation, the future development of ideas, and the support of the health care system and academia in Greenland.

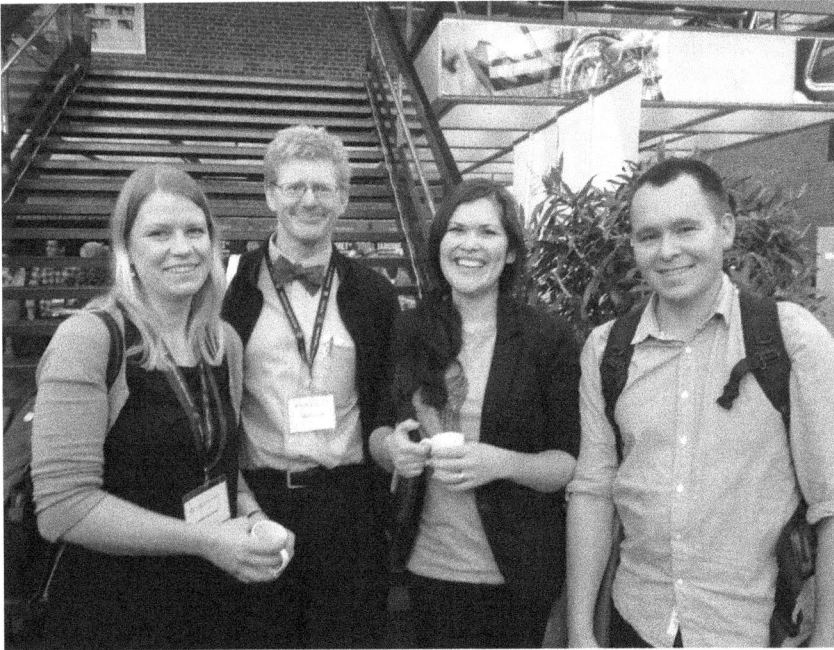

Figure 3.6 The Arctic Health research group at a medical convention

References

Andersen, S., Hvingel, B., and Laurberg, P. (1999) Graves' disease in the Inuit population of Greenland. *International Journal of Circumpolar Health*, 58: 248–253.

Andersen, S., Hvingel, B., Kleinschmidt, K., Jørgensen, T., and Laurberg, P. (2005). Changes in iodine excretion in 50–69-y-old denizens of an Arctic society in transition and iodine excretion as a biomarker of the frequency of consumption of traditional Inuit foods. *American Journal of Clinical Nutrition*, 81: 656–663.

Andersen, S., Jakobsen, A., Rex, H. L., Lyngaard, F., Kleist, I.-L., Kern, P., and Laurberg, P. (2013a). Vitamin D status in Greenland: Dermal and dietary donations. *International Journal of Circumpolar Health*, 72: 21225.

Andersen, S., Jakobsen, A., and Laurberg, P. (2013b). Vitamin D status in North Greenland is influenced by diet and season: Indicators of dermal 25-hydroxy vitamin D production north of the Arctic Circle. *British Journal of Nutrition*, 110: 50–57.

Fleischer, I., Schæbel, L. K., Albertsen, N., Sørensen, V. N., and Andersen, S. (2017). Diagnosis of osteoporosis in rural Arctic Greenland: A clinical case using plain chest radiography for secondary prevention and consideration of tools for primary prevention in remote areas. *Rural Remote Health*, 17: 3910.

Jakobsen, A., Laurberg, P., Vestergaard, P., and Andersen, S. (2013). Clinical risk factors for osteoporosis are common among elderly people in Nuuk, Greenland. *International Journal of Circumpolar Health*, 72: 19596.

Krarup, H., Andersen, S., Madsen, P. H., Okels, H., Hvingel, B. H., and Laurberg, P. (2008). Benign course of long-standing hepatitis B virus infection among Greenland Inuit? *Scandinavian Journal of Gastroenterology*, 43: 334–343.

Noahsen, P. and Andersen, S. (2013). Ethnicity influences BMI as evaluated from reported serum lipid values in Inuit and non-Inuit: raised upper limit of BMI in Inuit? *Ethnicity and Disease*, 23: 77–82.

Noahsen, P., Kleist, I., Larsen, H. M., and Andersen, S. (2019). Intake of seaweed as part of a single sushi meal, iodine excretion and thyroid function in euthyroid subjects: a randomized dinner study. *Journal of Endocrinological Investigations*, doi:10.1007/s40618–40019–01122–01126.

Rex, K. F., Krarup, H. B., Laurberg, P., and Andersen, S. (2012). Population-based comparative epidemiological survey of hepatitis B, D, and C among Inuit migrated to Denmark and in high endemic Greenland. *Scandinavian Journal of Gastroenterology*, 47: 692–701.

Schaebel, L. H., Vestergaard, H., Laurberg, P., Rathcke, C. N., and Andersen, S. (2013). Intake of traditional Inuit diet vary in parallel with inflammation as estimated from YKL-40 and hsCRP in Inuit and non-Inuit in Greenland. *Atherosclerosis*, 228: 496–501.

Schaebel, L. H., Bonefeld-Jørgensen, E. C., Laurberg, P., Vestergaard, H., and Andersen, S. (2015). Vitamin D-rich marine Inuit diet and markers of inflammation: A population-based survey in Greenland. *Journal of Nutritional Science*, 4(c40): 1–8.

Schaebel, L. H., Bonefeld-Jørgensen, E. C., Vestergaard, H., and Andersen, S. (2017). The influence of persistent organic pollutants in the traditional Inuit diet on markers of inflammation. *PLoS-ONE*, 12: e0177781.

Sørensen. V. N., Wojtek, P., Pedersen, D. S., and Andersen, S. (2015). An efficient case finding strategy to diagnose osteoporosis in a developing society with low treatment frequency. *Journal of Endocrinological Investigations*, 160: 4882–4885.

4 Arctic Auto-ethnography

Unsettling colonial research relations

Naja Dyrendom Graugaard

Introduction: Positioning in the suspense

In late January 2017 my partner, our two children, and I packed our bags, closed up our residence in Denmark, and left to spend seven months in Greenland. As a part of my PhD programme, I was going to co-teach at Ilisimatusarfik (University of Greenland) in Nuuk for three months. I had planned a subsequent four-month travel along the Greenlandic west coast, conversing with hunters and locals working in relation to seal hunting and sealskin production. These interviews were going to make up a considerable part of the "data collection" for my dissertation, which explores the relations between narratives of *puisit* [seals], Kalaallit[1] sealing practices, and decolonization processes in Greenland. I was therefore interested in having conversations about Kalaallit engagements with and relations to *puisit*, as I wished to inform my research with local and Indigenous perspectives on sealing and sealskins.

For my partner and our one-year-old son, it was their first stay in Greenland, and I looked forward to introducing them to my maternal relatives, whom we were going to visit along the way. Both of our children, who carry my *ningiu's* [grandmother's] and her younger brother's *atit* [names], would also (re)convene with the family relations that they have inherited through their namesakes. Besides my excitement for the journey, I was also nervous and anxious about stepping into the role of a researcher, snooping around for people's time and stories. Being a mixed-bag, *qallu-naaq-kalaaleq*[2], personal product of the colonial relations between Denmark and Greenland, my worries were fuelled by my inabilities in *kalaallisut* [the West Greenlandic Inuit language], my white coding, and the privileges that I inhabit from my middle-class upbringing in Denmark (see Graugaard, 2016). Positioning myself as an academic doing "fieldwork" in Greenland's coastal villages did not exactly balance out the existing power dynamics between myself, my fatherland, and my other motherland. It arguably entangled new power relations to navigate in my "borderland" transgressions (Anzaldua, 2012) between Denmark and Greenland.

Owing to these and various other issues related to entering Greenland as a "field of research," which I elaborate in the next section, I wished to decolonize my interview approach (inspired by, for example, Smith, 2012). To me, this involved breaking down the interviewer–informant dichotomy, and shifting the control over the process and outcome of the interview over to the participants. My ambition was to part with "hit-and-run" research (and other colonial forms of research) in Kalaallit communities. In retrospect, however, my fieldwork failed as much as (if not more than) it consolidated new steps toward decolonizing Arctic research. Always embedded in a maze of colonial entanglements, my interview encounters did not facilitate the decolonial and collaborative research spaces to which I had aspired. While I had planned for an interview process that was constituted by co-creating the terms, topics, and questions together with the informants, my conversations in Greenland usually ended up imitating conventional forms of semi-structured qualitative interviews (e.g., Kvale, 1996). This was probably due to many and various aspects: language barriers, time constraints, different expectations, relational uncertainties, and a lack of skills (on my part). Most of all, my presence by way of research seemed to recreate the power asymmetries which I sought to interrupt. In this light, the path to co-creation was perhaps far-fetched and possibly forced. It became ever more apparent to me that the efforts to decolonize neither come with a manual nor are they a "metaphor" to legitimize my research endeavor (Tuck and Yang, 2012).

Taking my (sense of) failure seriously, there might still be something to be learned from my fieldwork experiences. Agitated and disturbed by the colonial relations of which I was unavoidably a part of and often recreating, each research encounter instigated new *cycles of reflection*. Here, the "good times," the smiles and laughs, and the moments of (seemingly) mutual understanding in the interview sessions did not appear to me as reassuring comforts to *settle on*. Rather, working with my uncertainty and unease in the "data collection" process became a methodological point of departure and return. In this sense, my fieldwork revolved as a "borderland" (Anzaldua, 2012) of sorts, which risked *resettling* and *unsettling* colonial relations at one and the same time. Attentive to my (sense of) failure, my engagement with research participants constantly nudged me to check in, employ reflexivity, resist reproducing the colonial gaze, and strive for accountability. Importantly, this kind of engagement does not seek to evaluate the research encounters by distinguishing between "failure" and "success"—arguably, the process to decolonize does not rest on a language of (self-fulfilling) accomplishments (Tuck and Yang, 2012). Alternatively, sensing my failures to decolonize indicated the ongoing struggle, commitment, and obligations involved in unsettling research encounters, which inhabit multiple colonial histories and entanglements in their making. Employing an auto-ethnographic perspective enabled me to labor with my unease and shortcomings as a way to continuously disturb the coloniality of my research.

In this chapter, I thus reflect on my methodological learnings as an auto-ethnographic study. As this chapter meditates on the possibilities of breaking with neo-colonial traditions in Arctic research, I will first discuss the emic notion of hit-and-run research in Greenland, its historical legacies and contemporary consequences. Subsequently, I will discuss some of the theoretical reflections and methodological responses offered by auto-ethnographic studies, thereby situating my own fieldwork methodology. My call to employ more auto-reflexivity in Arctic research will be exemplified by a personal encounter in Nuuk, where I became an interviewee for a Swedish film documentary and the subject of a nonreciprocal research project *on* and *about* Greenland (cf. Smith, 2012). These reflections will be followed by three vignettes from my experiences as a fieldwork researcher in Greenland. The vignettes work through the relations between researcher and informant, researcher positionalities, and ways in which research can reproduce or resist colonial forms of research. By their example, the vignettes reflect (on) auto-reflexive processes which *potentially* challenge colonial self–other and researcher–researched relations. While a reflexive auto-ethnographic approach does not necessarily lend itself as "collaborative", I show how it can enable some of the footwork for more collaborative research methodologies in the Arctic. As scholars have been questioning the schemes of participatory and collaborative research in colonized, Indigenous communities (Zavala, 2013: 59), this auto-ethnographic "footwork" pushes new reflections on the very concept and meaning of participation and collaboration. Rather than offering a recipe for "best practice," I propose that engaging with Arctic research encounters through *cycles of reflection* is informative to reconceptualizing more ethical ways of researching with Indigenous communities.

Hit and run

In recent years, hit-and-run research has become a notable, negatively loaded term in Greenland; I have particularly noticed its use among Kalaallit university students. Here, the term refers to the tendency of international scientists to fly into Kalaallit communities or unpopulated areas (e.g. the ice cap), remaining only as long as it takes to collect their research data. This can be characterized as a fly-in/fly-out data extraction with little or no consolidation in the local or surrounding communities. In particular, hit-and-run research describes a nonreciprocal process of knowledge production which benefits the incoming scientist while neither exchanging with, reporting to, nor "paying back" locals or Greenland as a whole. The term has primarily been used to portray such tendencies in non-Kalaallit natural science projects (Hauptmann, 2016) but it can also be used to define some of the many qualitative social science projects in Kalaallit communities that share the same characteristics (own observations).

Arguably, the Greenlandic pinpointing of hit-and-run research does not occur in a vacuum; it echoes the history of colonial exploration and exploitation in and of Indigenous communities and peoples in the Arctic. It associates how Inuit peoples and ways of life have been described and defined, classified and collected, and appropriated and represented by polar explorers, colonists, scientists, and anthropologists in, with, and throughout the colonization of the Arctic (Krupnik, 2016; Nungak, 2006; Petterson, 2012). Since the mid-16th century, Europeans have conducted research and writings on and about Inuit. According to Igor Krupnik's (2016: 1–2) historical overview, the first studies on "Eskimos" constituted a major European enterprise, which was subordinate to the interests of the mission and Christian conversion, colonial expansion and trade, and journeys of "discovery". Thuesen et al. (2017: 102), specifically, state that the knowledge production on Greenland—initiated by missionaries and employees in Denmark's Royal trade department—was financed by the Danish state as an act toward establishing and claiming Greenland as a Danish colony. These first studies on the Arctic laid the foundations for Inuit studies. They were further developed with the pursuit of "scientific knowledge" in the 19th and 20th centuries, when researchers from various disciplines started undertaking more systematic, standardized studies on Arctic Indigenous peoples. They made ethnological collections and documentations, recorded physical and body measurements, excavated ancient sites and skeletal remains, and created cultural typologies (Krupnik, 2016: 3–10). As Krupnik (2016: 3–6) illustrates, this new scholarship on "Eskimos" was dominated by white European and American men who were influenced by traditions of naturalism, typology, evolution, and environment-induced development. As is furthermore pointed out in the work of Christina Petterson (2012), these studies also embodied Western anthropological racism. Drawing on the work carried out by Ole Høiris (1983), Petterson (2012: 31–32) argues that Danish research in Greenland provided empirical data for Denmark's participation in racial studies that started in the late 19th century and continued until World War II. In and of itself, the unrestricted access enjoyed by Western scientists to Arctic communities in which they could make Indigenous peoples their forced subjects of research, was a premise and product of white supremacy (Tuck, 2009: 412).

Thus, undertaking research in Kalaallit communities cannot avoid being associated with European (particularly Danish) imperialism and colonialism. Similar to Indigenous experiences with research across the Arctic and worldwide, "scientific research" can conjure up bad memories: measurements of skulls and intelligence, unsanctioned experiments, camera lenses, erroneous ethnographies, knowledge theft, empty promises, and so on (Battiste, 2008; Kuokkanen, 2008; Smith, 2012; Tuck, 2009; Tuck and Fine, 2007; Wilson, 2008). Being a "dirty" word in the experiences of many Indigenous communities, "research" has come to represent the colonial and racializing practices by which knowledge *about* and *on* Indigenous peoples has been collected, defined and represented back into Western societies (Smith, 2012). However, Indigenous peoples and cultures have been and still are frequent "objects" of research. As Opaskwayak Cree scholar Shawn Wilson (2008) comments on how often "this research has neither been asked for, nor has it had any relevance for the communities being studied" (p. 15). The formulations of hit-and-run research in Greenland arguably articulate a similar experience.

Nonetheless, Krupnik (2016) argues that the discipline of Inuit studies underwent a major, decades-long transformation beginning in the 1970s, which addressed some of the colonial legacies in Arctic research. In his words, the transition from being a discipline of *Eskimology*[3] to becoming *Inuit Studies*, signified a change from conducting research *about* Inuit, to being research *of* and *for* Inuit (2016: 14–18). This is reflected in the establishment of Ilisimatusarfik (University of Greenland) and the growing contributions of Kalaallit scholars[4] who have influenced the scopes of Greenlandic studies. Yet, some Inuit (including Kalaallit) scholars and Arctic residents remain critical of the said transformation and call for its actual implementation (e.g., Markussen, 2017; Nungak, 2006; Rasmussen, 2002 and 2004; Pfeifer, 2018; Williamson, 2014). In practice, most authors of Arctic scholarship are still academics from outside of the Arctic who produce works depicting and discussing different dimensions of Arctic living and culture, and this tendency might not diminish in the face of increasing global interest in Arctic matters. All things considered, when browsing the contemporary scholarly works on Kalaallit and Kalaallit life forms, it does not take long to conclude that most of it is written by Danes. While the growing interdisciplinary body of Arctic scholarship is displaying increasing levels of ethical sensitivity and collaborative approaches (as is illustrated in this volume), I also argue that many Arctic scholarly works continue to underscore a research process in which the academic author features as the expert-researcher, who studies and describes Arctic peoples in terms of the research*ed*—and leaves the researched with little or no control over the process and outcome.

Current Kalaallit articulations of hit-and-run research can thus be placed in the continuum of unequal relationships, which the "studying the Arctic" discipline already inhabits. Similar to Wilson's aforementioned characterization of incoming research in Indigenous communities (2008: 15), the researcher who "hits and runs" in Greenland neither asks permission nor waits for a response; nor does he return what he has borrowed, taken, or learned. While an auto-ethnographic approach does not directly resolve these problematics, I will discuss how greater auto-reflexivity as an Arctic researcher can provide ways to *resist* or *refuse* hitting and running.

Paths to auto-ethnography

Auto-reflexive qualitative research has been described as "turning back on yourself the lens through which you interpret the world" (Goodall, 2000 cf. in

Mainsah and Prøitz, 2015: 171). Employing reflexivity in research thus intends to reveal and enunciate the connections between the writer and her own subject as a way to reflect on the lived experiences which are studied (Goodall, 2000). In the growth of this scholarship, auto-ethnography has emerged as a methodology that "facilitates reflexivity about researcher positionings and the politics of locations" (Lapina, 2017: 5). By considering positionalities and relations within the research, an auto-ethnographic approach urges the researcher to reverse "the gaze". This entails acknowledging the baggage of her background and accounting for how interactions and encounters with research participants also shape and structure texts, arguments, and explanations (Tomaselli et al., 2008: 354). In such ways, auto-ethnography offers a challenge to and redefinition of the traditional fields of ethnography and anthropology, which are disciplines that have been deeply involved in manifesting the colonial gaze and the representation of the indigenous Other (Smith, 2012). Rather than collecting and describing "the exotic other," an auto-ethnographic researcher is reflective of her own life, her cultural practices, and social assumptions as being essential to understanding the research encounters (Tomaselli et al., 2008). In such ways, reflexivity reflects a dedication to destabilize convictions of "omnipotent expertise" and presents a critique of the status of knowledge and representation within qualitative research.

An auto-ethnographic approach thus considers the subjective location of the researcher as being conditional to knowledge production (Ellis and Bochner, 2000). In Linda Lapina's (2017: 60) auto-ethnographic work, researcher locations and positionalities are described as *shifting* and *changing* rather than fixed in categories of, for example, *white, woman, privileged*. However, they may still be located in the complex structures of colonial relations (Nicholls, 2009: 118). Mediated intersectionally, scholar positionality also entangles the broader political geographies of communities, norms, and institutions. Elaborating on Donna Haraway's (1988) approach to knowledge as "situated" and "partial," Lapina (2017: 60) holds that research encounters are furthermore grounded in places and/as bodies. Similarly, Mainsah and Prøitz (2015) approach the body of the researcher as a (privileged) site of knowledge production, in order to illustrate how living and theorizing produce and structure each other. Here, senses and feelings are foregrounded to provoke new perspectives, and to account for "what we sense and what we create as academics is a process of human encounters, movement and entangling" (p. 170). As feminist scholars argue, the feelings and emotional entanglements between researcher and researched are co-constructive of a research encounter (Lapina, 2017: 60). In order to highlight and expose these interactional textures of undertaking research, auto-ethnographies often utilize narratives and anecdotes, writing in first person, vignettes, and memory work. It also invites an inclusion of the research *process* as a topic in and of itself, so that contradictions of academic discourse and knowledge production are exposed (Lapina, 2017; Mainsah and Prøitz, 2015; Tomaselli et al., 2008: 354).

By "turning back the lens" on the Arctic researcher, it thus becomes possible to recognize her subjective locations as being conditional to the knowledge production on the Arctic and to destabilize the convictions of omnipotent scholarly "expertise". As my position as an Arctic scholar is often located in the in-between-ness of "the borderland," as well as in the privileges of whiteness—while shifting and changing, intersectionally (Lapina, 2017: 60)—my auto-ethnographic approach urges me to "reverse the gaze" and account for how my baggage and backgrounds shape inter-actions and research encounters. In this sense, an auto-ethnographic approach also challenges the tradition of Inuit studies and its legacy of gazing on the Arctic and constructing the "Eskimo other" (Fienup-Riordan, 1990; Krupnik, 2016). Arguably, an Arctic auto-ethnography should not simply work to clarify positions as a kind of "hand-wringing, the flash of positional confession before proceeding as usual" (Tuck and Yang, 2014: 814). Instead, I hold that an auto-ethnographic approach can seek to employ a kind of reflexivity which is responsive to research participants and other aspects of a research endeavor, such as its terms, topic, process, and outcomes. Nicholls has argued for a "multi-layered reflexivity" which is also committed to the inter-personal, community relations and the possibility of "ceding researcher control" (Nicholls, 2009). In my work, I hold that recognizing and accounting for the unease, uncertainty and (sense of) failure which accompany asymmetrical research relations is one way toward "ceding" or unsettling researcher control. Working auto-ethnographically then fuels the cycles of reflection that hold me to continue unsettling colo-nial relations, rather than settling on them. At times, this might require "blocking the gaze" and practicing "refusal" in Arctic qualitative research (Simpson, 2007; Tuck and Yang, 2014), rather than merely "turning the lens". *Refusal* is described as an objection to making someone or something the subject of research, and a form of un-claiming, objectless analysis. This involves an active resistance to undertaking damage-centered research (also critiqued in Tuck, 2009; Simpson, 2007; Tuck and Yang, 2014), which trades in stories of pain and humiliation of communities and peoples. It thereby not only reverses but also *blocks* "the settler colonial gaze that wants those stories" (Tuck and Yang, 2014: 812).

In the next sections, I engage auto-ethnography as a way of critically reflecting on and challenging the research relations in which I have partici-pated as part of my fieldwork, through *vignetting* and *re-membering* from "the field". The vignettes are "performative narratives" which draw on memory work, journal entries, interviews, conversations, fieldwork experi-ences, and personal reflections. Writing the auto-ethnographic vignettes then presents a new cycle of reflection, which can result in "deeper knowing" (Mainsah and Prøitz, 2015: 184). While sensing my failure to destabilize power asymmetries and the colonial gaze in my fieldwork, writing about it auto-ethnographically provides new opportunities to do so. While my auto-ethnography does not emerge as a mere solution to the colonial aspects of undertaking Arctic research, it can by example suggest ways to practice more responsive, affective, and ethical research. In order to situate this call for auto-reflexivity, I first "re-member" (Graugaard, 2016: 19) an experience of *becoming an interviewee*.

Turning the lens—becoming the interviewee

While living in Nuuk, a new friend asked me to participate in his Swedish-produced documentary. Our friend was (and still is) a curious, warm-hearted, and well-intended film documentarist, who wished to interview me about the historical, contemporary, and personal relations between Denmark and Greenland. I agreed. Sitting on the other side of the interviewing table and becoming the object of a camera lens proved a rather insightful experience. I experienced first-hand the effects of an interview's terms and framework; I learned how knowledge, emerging from an interview, is co-produced in the relation between the interviewer and the interviewee; and I had a very visceral experience of being gazed at. Besides being an encounter that created discomforts and personal disturbance, it also provided new critical reflections on my own interviewing practice.

My friend described his style as "seeing things, the way they are" and "in natural surroundings". Therefore, he did most of the filming and interviewing in our apartment, over coffee and cake with our intimate families present, and with a backdrop of noisy, playing kids. He filmed, it seemed, whenever he thought someone said or did something that he considered useful to his project. This left me feeling nervous and tense, as I sensed the expectations behind the camera. At times, I felt as if he lost interest in what I was sharing if it was not "dramatic" enough. As the interviews unfolded, the questions started to circle around "problems," being suggestive of "identity struggles". Subject to an increasingly "damage-centered" lens (Tuck, 2009; Tuck and Yang, 2014), I began to feel as though I was expected to fulfil a preconceived role—and that my agency of breaking out of it was limited. After each of the three interviews, I felt overwhelmed and exhausted. I could not really remember how I had responded to the questions, and doubts started surfacing: *What had I actually said and told? What happens when my responses are digitalized on video? What if my perspective changes in a couple of months—or even today?* It felt disturbing to me that the circumstantial stories emerging from these encounters could be represented as evidential depictions of Greenland and received as durable ethnographic documentation about Kalaallit life forms.

This interviewing process left no room for rewinding, revisiting, reconsidering, or retaking. I was not invited nor allowed to partake in (de)selecting the interview parts which I considered (un)representative of my perspective and experience. The limitations to my influence on the terms and process became particularly clear to me during our final interview. After a full day of teaching and low on energy, I felt weary and annoyed when my friend behind the camera posed yet another question about my "in-between identity".

Therefore, I asked if we could take a short break, and I hoped to get a chance and courage to voice my concerns about his questions. My friend agreed to a break, but he did not turn off the camera. Being continuously recorded and recitable, I felt trapped and speechless in the light of the filming camera. While my friend conceived his methodological approach as part of observing and documenting the raw material of "reality"—"seeing things, the way they are"—I perceived it as a reality that was as much produced by the various conditions of its setting. Being positioned as an "informant," I recognized that my responses and statements were shaped by multiple dimensions of the interview, such as how the way the questions were phrased and posed, my (in) ability to influence the conversation topic and course of events, our emotional states of being, and the relation and power dynamics between myself and the interviewer (cf. Smith, 2012; Nicholls, 2007; Zavala, 2013). The "reality" that the interview was reflecting was particularly dependent on my friend's gaze and my stifled attempts at stepping into or out of it.

Becoming a subject of investigation in my friend's documentary had offered me a sensory, visceral kind of reflexivity, which further urged me to scrutinize the conditions of my own research. It no longer seemed fitting to approach my "data collection" as raw materials of reality: A researcher's embodied self also produces the realities which she experiences (Mainsah and Prøitz 2015: 183). My concerns, which arose while being an "informant", supported my ambitions to *co-produce* the interview questions with the interviewees, to make my research objectives *transparent*, and to create an interviewing process which was available to *changing*, rewinding, retaking, and deleting. I therefore decided to strive to meet the interview participants more than once (preferably two or three times), if doing so was possible and agreeable. I also wished to make the interviews available to each participant, and to give them an opportunity to edit my presentations and analyses. The film documenting experience also brought home the importance of consistently checking my positionings, actions, and assumptions as an interviewer (Tomaselli et al., 2008).

The following vignettes only bring forth *some* and *parts* of these reflections as they arise in my fieldwork: in-situ and partial, grounded in different places and embodiments (Haraway, 1988; Lapina, 2017). My vignettes work through my unease with and (senses) of failures to destabilize power asymmetries as a way of engaging in new cycles of reflection on methodological practice. They are thus not stories of "successful" research encounters and fulfilled ambitions. Instead, the vignettes provide a basis upon which to meditate on how to resist and unsettle (neo)colonial relations when undertaking research in the Arctic. Inspired by the work of Mainsah and Prøitz (2015), I provide small post-scripts after each vignette in order to elaborate on the new reflections and challenges that each experience brought to my fieldwork. Engaging with the informants and our conversations auto-ethnographically, I have aspired to a non-objectifying style of writing. Resisting to search for the subjectivity of the Other (Tuck and Yang, 2014: 815), I seek to engage the encounters in new cycles of reflection on my research practice and methodology.

Places and spaces of power in the interview

On my walk over to Salomine, I think about my aspirations to decolonize the research interview. I feel as though I have created an oxymoron by attempting to stick a decolonial label on a colonial undertaking of doing ethnographic fieldwork in Kalaallit communities.

No matter how I twist and turn it, I am still on fieldwork and scouting for people who can inform my project. Travelling the west coast of Greenland to "search for" local Indigenous perspectives, I see myself hypocritically reproducing what I wish to break down.

Ultimately, I am a researcher on my way over to interview Salomine about her views on sealskin, her sewing skills, and her life as a hunter's wife.

I repeat to myself that I am dedicated to change things around and to resist assuming control over the situation as "the expert". Determined to co-produce the interview questions, I wish to leave it to Salomine to decide the scope and extent of our conversation. I wish to converse on her terms—not my own.

It all turns out to be trickier than I had expected, and my pre-conceived plan for "decolonizing" the interview falls through.

I talk a lot in the beginning of our meeting in order to introduce my method, my project, and my purpose. My intention is to provide more information about my research, before Salomine agrees or disagrees to participate. Yet do I instead just appear more "expert-like" and with less presence?

Focused on following my plan to co-produce the interview questions, I am slow to respond to the confusion, awkwardness, reluctance, and linguistic barriers that my ideas are creating.

Eventually, I pose some of my pre-prepared questions from my interview guide to move beyond silence and to amend the situation. Altogether, I must have appeared like a rather unknowledgeable, unprepared researcher.

Toward the end of this first meeting, Salomine asks if I can make some questions for our next meeting, instead of creating them together.

Post-script: This encounter reminds me of the statement made by Eve Tuck and Wayne Yang (2012) that "decolonization is not a metaphor" (p. 1), and that it does not come in neat, pre-prepared, labelled packages. In this first meeting with Salomine, I was so focused on fulfilling my own preconceived ideas about what a more ethical, collaborative, decolonial interview would be and look like that I forgot to be present in establishing our relationship in the interview. My concern with tackling the potential power asymmetries in the interview probably resulted in an excessive concern with my researcher-self. In effect, my preoccupation with my own positioning obstructed my ability to, affectively, apprehend Salomine's position. In such ways, the efforts to manage subjectivity in reflexive work can result in "emphasizing the distance between self and others" (Nicholls, 2009: 122). It might have been better to begin by checking with Salomine and asking what kind of interview *she* would prefer?

Thinking that I was reversing my researcher gaze, I had yet to reflect on my own assumptions. My attempt to break down power asymmetries was based on the assumption that Salomine needed and benefited from my specific approach (Tuck and Yang, 2014: 813). Meanwhile, I assumed that my own position was perceived as one of power and expertise (aside from my occasional worry that I appeared "unprepared" and "unknowledgable"). Thus, in my framework of "conversing on her terms," I had not considered what Salomine's terms actually were or would be. She had to remind me toward the end of our meeting: "Can't we do it in a way where *you* make some questions?" Ironically, my efforts to rid our interview of colonial self–other relations may instead have reinforced them. Even prior to our first meeting, I assumed to know what such relations looked like, what they were about, and what it would take to resist them. Busy with resisting my researcher position, I had re-centered myself as the one with the answers to overcoming the problematic, without asking the participant for whom it mattered. In order to work away from *controlling* collaboration, Nicholls therefore underlines the importance of "attention and receptivity to the relational... spaces" in reflexive work. She also argues that working reflexively does not demand a denial of researcher subjectivity, but rather an acknowledgement of this role—and the need to build relationships of trust in that acknowledgement (2009: 122–23).

It's good that you came here–or is it?

The conversations with Ruth and Pavia touch me deeply.

I am in the village of my maternal family, but a pressing deadline has lulled me into a world of books, articles, and writings. My meeting with Ruth and Pavia gently bursts my bubble.

Pavia tells me that our ancestors used to "sail together" (which involves hunting)—my great-grandfather and his father, and their sons. We are connected like that. This is news to me. Our connection makes me feel happy and emotional. We talk about how the writing of Greenland's history often seem estranged from lived experiences.

Ruth and Pavia show me an old magazine with a story mentioning my grandmother's graduation from the teacher's college in Nuuk. We talk about Lulu, about Nathan, and about Karen Marie. We also talk about Aapi and Atsa (my children) and their home-coming to Arsuk.

It does not feel like a meeting with a stranger over a cup of coffee and attempting to read each other. Our ancestors had sailed together.

Our meeting reminds me of conversations with my grandmother. I sit down and listen. Pavia and Ruth tell me what they wish to share and what they think is relevant for the topic. I never bring out my question sheet.

At the end of our conversation, Malene (their niece and our translator) says: "It's good that you came here."

I am surprised, mixed with feelings of doubt and self-assurance. Did the interview mean something to them?

My tummy ache disappears, and I am not embarrassed about the whole situation of interviewing and asking questions like I usually am.

Post-script: Undertaking research in my family village seemed to hit a core note in my methodological approach. Here, the intimate connections between myself, "informants", and the fieldwork appeared to be essential conditions of my research. As personal relations and ancestral ties surfaced (as they often do in small communities like Greenland), it neither made sense nor was it possible to endorse a sense of anonymity. In very practical terms, auto-ethnography's objection to viewing research as "an unlocated stable reality" (Mainsah and Prøitz, 2015: 171) was brought to light: I was indeed located, and my interview with Ruth and Pavia sprung from our crossroads. For this reason, this research interview was highlighted as a personable encounter in which our interrelations were necessarily implicated. While this could position my research as "un-objective" in scientific, positivist terms, it supports the notion that a research encounter is always grounded in places and/as bodies (Lapina, 2017: 60). My conversations with Ruth and Pavia were particularly situated by our specific place-based and ancestral relations. These layers of shared familiarity allowed for interviews that were not dependent on scripts or interview guides. Prior to the meeting, a mutual friend had informed Ruth and Pavia about my project, which gave them some time to reflect and prepare. This seemed to be reflected in our conversation. They seemed to have thought about what they wished to share with me, and in doing so, it was primarily up to them to determine the scope of our conversation. With my grandmother's name on a sheet of paper on their coffee table, it felt less like an interview and more as if they were entrusting me with their stories.

It is tempting to read this encounter as my license to undertake fieldwork in my family's village and to infuse my research with self-importance and legitimacy. Seeking a quick cure to my tummy-aching fear of being a researcher who "hits and runs," the statement "it's good that you came here" can feel reassuring. Rather than merely indulging the comforts of feeling accepted, however, the experience also unsettled me and impelled new cycles of reflections, which questioned: *Was it really good that I was there? In what ways? Why? For whom?* Bound by relation, these questions still seem particularly urgent. Research implicates our relations. I have engaged people's time and knowledge; I have entered homes, met families, and enjoyed coffee; I have asked for help and direction. Overwhelmed by all that has been given to me in the name of my research, I still seek out ways to engage with it in reciprocal and responsive ways. Once transcribed and digitalized, processed, and captivated in new forms and shapes, my research can quickly become estranged from the lives that it includes and involves. Like writing this chapter: How do I communicate the voices of those who have spoken to me on paper? And without stripping them of their own agency? How do I carry the stories responsibly? And consider the repercussions and implications of retelling them? Furthermore, how do I know when my retelling is asked for and when it is refused? Similar questions have occupied Indigenous and non-Indigenous scholars who undertake research in colonized, marginalized, Indigenous or other minority communities (Kuokkanen, 2008; Nicholls, 2009; Simpson, 2007; Smith, 2012; Wilson, 2008). Wilson (2008) responds with the concept of "relational account-ability," which suggests that a research methodology should be based in a community context and demonstrate respect, reciprocity, and responsibility when put into practice (p. 99). Other scholars have added, subtracted, and elaborated, but do not provide solid solutions—which is not necessarily the intention, either. I do not think that there is a formula; the questions must be considered every time and in each case. Acknowledging that research implicates our relations (whether they are ancestral, new, or yet unknown) is not a license to undertake research, but perhaps more a dedication to keep asking those questions and seek the answers.

Distorting the interviewing voice

> Seated around the kitchen table, I give Erik my sheet with questions, but I tell him that we don't have to follow it. I can't remember if he ever looks at them.
>
> I ask him some introductory questions. The conversations flows—and at times it stumbles. It is hard for Erik to hear me, and Tina must come to help.
>
> Erik shares many things. In our conversation, we are in East Greenland and out hunting, and we are at the hooded seal breeding grounds that were over-hunted by Norway. We also touch on some of the ancient stories, which used to guide the ways of hunting. But mostly we talk contemporary seal hunting politics.
>
> I sense it's time to wrap things up. I ask a couple of clarifying questions, and we get to talking again.
>
> And two hours have suddenly passed by since we started. Erik is tired now. I feel bad for tiring him.
>
> He says that he hopes that I can use his answers. He says that he doesn't hear that well. He says that he can't hear what I say most of the time. That he guesses what my questions are...
>
> Oh!?
>
> On my walk back, I'm thinking about how thrilled I am about Erik's last comments: He guessed my questions most of the time.

Post-script: Since my meeting with Erik, I have often entertained the thought of introducing a kind of distortion device in future interviews. A "Darth Vader device" of sorts (perhaps more friendly-looking) which could obscure my voice a bit, thereby making my questions a bit more ambiguous, slightly less certain, and a bit more open to interpretation. Even though the auditory challenges might have obscured my interview with Erik, they also opened it up in unforeseen ways. Guessing my questions most of the time, Erik must have had a good deal of decision-power in terms of what and how to respond to my questions. At the same time, I became less vocal, as I experienced the efforts involved in hearing my speaking. Perhaps the inaudible disruptions of my interview inquiry worked as small "gestures of inclusion"; by decentering my voice, they may have opened new grounds for Erik to negotiate the course of the interview (Nicholls, 2009: 124). In this unplanned-for manner, the terms of our encounter had subtly shifted, and our positions of inquirer vs inquired had become somewhat unsettled. At least and if nothing else, the encounter exemplifies a challenge to the researcher's "god-gaze of the objective knower" (Tuck and Yang, 2014: 815)—it was only by the grace of Erik that I became aware that my inquiries had been close-to-inaudible and guessable.

Noticeably, our conversation was characterized by stumbles, silences, and pauses, as much as by free flow. At times, I was uncertain about whether I had understood what was communicated, or whether I had nodded or laughed with appropriate timing. I also left the interview session feeling worried that I had gone overboard, stayed too long, and missed a hint to leave earlier. On reflection, our mutual "guessing" could have impelled further effort toward understanding one another. It could be viewed as a moving into the "liminal, in-between space," which is held together by "fragile and fluid networks of connections and gaps" but which is yet a possibility to collaborate (Nagar, 2009: 359, in Nicholls, 2009: 121). According to Nicholls (2009: 124), the possibility to collaborate and engage meaningfully requires a researcher's willingness to decenter herself and cede researcher control. This shift to researcher positionality precedes being receptive and holding space for new cultural domains of understanding.

Conclusion: Moving into the liminal, the in-between

Contemporary practices of hit-and-run research in Greenland reverberate with the colonial legacies of exploiting Arctic Indigenous peoples and lands in the name of research. Histories of unrestricted scientific access, the forced subjectification of local populations to experiments, and the objectifications of "Eskimos" to the white male gaze all characterize the birth of the discipline of Inuit studies. These histories stick to the natural or social scientist, flying in and out of the Arctic; who hits and runs. Current Arctic critiques of such non-participatory and non-reciprocal research endeavors call for a shift in practice.

The ways to shifting practice, within Arctic research, are diversely defined. Some scholars have termed it a transformation toward locally oriented research: by and for Inuit (Markussen, 2017: 309; Pfeifer, 2018: 29; Williamson, 2014). Others have encouraged more "qallunologies," which could be described as a kind of Arctic whiteness studies (Nungak, 2006; Rasmussen, 2002, 2004). Some scholars and residents have begun redefining the terms of Arctic research in collaborative, participatory frameworks (e.g., Krupnik et al., 2010; this volume). I discuss how shifting practices to create conditions for reciprocity (or collaboration) can also occur in the processes of self-reflection, which move us to change our own praxis. Employing an auto-reflexive conceptual approach can provide some tools to address (neo) colonial hit-and-run dynamics in the research, which we create and participate in as Arctic researchers.

In my own research in Greenland, my positionalities are often located as mixed, privileged, woman, white-but-not-quite, Danish-speaking, and incoming but sometimes home-coming. These positionalities can sometimes shift and at other times feel stagnant. However, I hold that "researcher positionality" in Greenland is always located in a field of multiple colonial entanglements, owing to its particular historical legacies and associations. Thus, I argue that researching in the Arctic cannot entirely avoid its coloniality, no matter which position(s) in the (complex) relations between colonizer and colonized one might find oneself. Ulunnguaq Markussen's (2017) work relates to this by stating: "Involvement of Arctic peoples in Arctic research is a necessary but insufficient condition for truly locally oriented research: The framework, the expectations, and design of the research must itself be decolonised" (p. 309). I claim that collaboration and decolonization in Arctic research also entail a consideration of researcher positionality and her relations. While I do not claim that auto-ethnography makes for a decolonial approach, it can provide reflections on ways to destabilize (our) (neo)colonial research practices.

In its conceptual and methodological framework, auto-ethnography can encourage the Arctic researcher to "reverse the gaze" and critically reflect on how positionalities and interactions shape the research. Instead of collecting "the Eskimo other," an Arctic auto-ethnography employs reflexivity about the researcher's own life, practices, and assumptions; these are recognized as conditional to the unfolding and understanding of research encounters. By paying attention to how knowledge is being co-constructed and co-produced, an auto-ethnographical approach poses a challenge to the status of Arctic knowledge, representation, and general ideas of "objectivity" in qualitative research.

Inspired by Linda Lapina's (2017: 60) elaboration of Donna Haraway's work (1988), I have specifically approached my research encounters as grounded in places and/as bodies, and as partial, situated knowledges. My vignettes provide a way to work through some of my fieldwork encounters in Greenland as they appeared in-situ, partial, grounded, and embodied in different places. In these vignettes, I have reflected on the power dynamics between researcher and researched, and self and other, in my own work—continuously seeking ways to shift from *resettling* colonial relations to *unsettling* them. My vignettes reflect neither exemplary fieldwork nor stories of success; making claims on "successful research encounters" would arguably work against the dedication to and hard work of decolonization. The point of analytical departure is, rather, my (sense of) failed aspirations. Engaging with them auto-ethnographically provides new cycles of reflections and opportunities to challenge and further my methodology:

Re-membering my meeting with Salomine revealed to me the hidden difficulties of "reversing the gaze." Busy with attempts at decentering my researcher-self, I had not checked my own assumptions on the meaning of "decentering," and I was unable to invite Salomine's perspectives on it. I learned that pre-occupations with *the self* quickly and subtly take the focus from *the relational*. My conversations with Ruth and Pavia brought in the "relationality" of research in a different way. Connected by place-based and ancestral ties, our meeting highlighted how research interviews always implicate the personal, inter-personal, relational, and communal. Yet relation is neither a licence to research nor a fixed accomplishment. Instead, the sense of relationality seemed to return to the questions: How do I hold my research accountable, reciprocal, and responsive to the people it has involved? In the third and final vignette, I reflect on my meeting with Erik and ponder on our auditory challenges as a method to decenter the researcher's voice and cede researcher control. This might have worked in small ways to shift our positionalities and motivated shared efforts to hear the other and listen for the words that could have been missed.

If there is something general to be learned from this auto-ethnographical engagement with my fieldwork, it may be the recognition that the research encounter is neither a stable nor fixed, nor calculable space. It involves senses and sense-making, movement, and entanglement (Mainsah and Prøitz, 2015: 170). I have experienced it as a space that presents multiple risks, some of which are unknown and unpredictable. Recalling colonial legacies, I approach the Arctic research encounter as a space inhabiting multiple relations of power and entanglements; nobody arrives there uninvolved and virtuous, or fully relatable and accessible. An auto-ethnographical approach urges researchers to practice reversing the gaze and questioning positionalities, as a (pre)condition to meeting, conversing, and apprehending the positions of participants. I suggest that this approach can specifically be guided by (senses of) failure rather than stories (or manuals) of success and compel a sticking-to-the-struggle, which unsettles new cycles of reflections on current practices. This chapter has discussed the potential of auto-ethnography to take the responsibilities of Arctic research seriously, to inhabit the space of uncertainty and failure, to uphold accountability, and to be "willing to move into liminal, in-between spaces, decentering [oneself] by challenging traditional notions of objective control between researcher and research participants" (Nicholls, 2009: 121), as well as to be willing to move out again when asked to.

Notes

1 *Kalaaleq* (*Kalaallit,* pl.) is the local-specific term for Greenlander—a term usually used synonymously with being Greenlandic Inuit.
2 *Qallunaaq* (*Qallunaat,* pl.) is the term for a Dane or non-Inuit, in Greenlandic (and other Inuit languages).
3 Somewhat ironically, the Department of Eskimology at the University of Copenhagen first changed its name in 2019. Its new name is Greenlandic and Arctic Studies (Hansen, 2019).
4 Notably, scholars such as Robert Petersen, H. C. Petersen, Inge Kleivan, Gitte Adler Reimer, Mariekathrine Poppel, Minik Rosing, and Lene Kielsen Holm.

References

Anzaldúa, Gloria (1987/2012). *Borderlands: The new mestiza = La frontera* (4th ed.). San Francisco, CA: Aunt Lute Books.

Battiste, Marie (2008). Research ethics for protecting Indigenous knowledge and heritage: Institutional and researcher responsibilities. In N. K. Denzin, Y. S. Lincoln, and L. T. Smith (Eds.), *Handbook of Critical and Indigenous Methodologies* (497–510). Thousand Oaks, CA: SAGE Publications.

Ellis, Carolyn and Bochner, Art P. (2000). Autoethnography, Personal Narrative, and Personal Reflexivity. In Denzin, N. K., and Lincoln, Y. S. (Eds.), *Handbook of Qualitative Research* (733–768). Thousand Oaks, CA: SAGE Publications.

Fienup-Riordan, Ann (1990). *Eskimo Essays: Yu'pik Lives and How We See Them.* New Brunswick, NJ: Rutgers University Press.

Goodall, H. Lloyd (2000). *Writing the New Ethnography (Vol. 7).* Lanham, MA: Altamira.

Graugaard, Naja D. (2016). Uanga ("I"): Journey of Raven and the Revival of the Spirit of Whale. *KULT – Postkolonial Temaserie*, 14: 6–22.

Hansen, Jesper (2019). Navneskift: Slut med eskimologi. *Sermitsiaq,* January 28. Retrieved from https://sermitsiaq.ag/node/211175.

Haraway, Donna (1988). Situated knowledges: The science question in feminism and the privilege of partial perspective. *Feminist Studies*, 14(3): 575–599.

Hauptmann, Aviaja L. (2016). Hit-and-run forskning i Grønland og forskningsetik. *Ingeniøren*, October 24. Retrieved from https://ing.dk/blog/hit-and-run-forskning-groenland-forskningsetik-187727.

Høiris, Ole (1983). Grønlænderne i dansk antropologi før 2. verdenskrig. *Tidsskriftet Grønland*, 1: 30–46.

Krupnik, Igor (2016). From Boas to Burch: Eskimology Transitions. In I. Krupnik (Ed.), *Early Inuit Studies: Themes and Transitions, 1850s-1980s* (1–30). Washington, DCSmithsonian Institution Scholarly Press.

Krupnik, Igor, Aporta, Claudio, Gearheard, Shari, Laidler, Gita J., Holm, Lene K. (2010). *SIKU: Knowing our ice: Documenting Inuit Sea Ice Knowledge and Use.* London: Springer.

Kuokkanen, Rauna (2008). From research as colonialism to reclaiming autonomy: Toward a research ethics framework in Sápmi, presented at Kárášjohka, Norway, [November 23–24 2006] *Ethics in Sámi and Indigenous Research*, Report 1/2008: 48–63.

Kvale, Steinar (1996). *InterViews: An Introduction to Qualitative Research.* Thousand Oaks, CA: SAGE Publications.

Lapina, Linda (2017). Recruited into Danishness? Affective autoethnography of passing as Danish. *European Journal of Women's Studies*, 25(1): 1–15.

Mainsah, Henry andPrøitz, Lin (2015). Two Journeys into Research on Difference in a Nordic Context: A Collaborative Auto-Ethnography. In R. Andreassen and K. Vitus (Eds.), *Affectivity and Race: Studies from Nordic Contexts* (167–186). UK/USA: Ashgate.

Markussen, Ulunnguaq (2017). Towards an Arctic Awakening: Neocolonialism, Sustainable Development, Emancipatory Research, Collective Action, and Arctic Regional Policymaking. In K. Latola, H. Savela (Eds.), *The Interconnected Arctic – Uarctic Congress 2016* (305–311). Springer Polar Sciences.

Nicholls, Ruth (2009). Research and Indigenous participation: critical reflexive methods. *International Journal of Social Research Methodology*, 12(2): 117–126.

Nungak, Zebedee (2006). Introducing the science of Qallunology. *Windspeaker Publication*, 24(2).

Petterson, Christina (2012). Colonialism, Racism and Exceptionalism. In K. Loftsdottir and L. Jensen (Eds.), *Whiteness and Postcolonialism in the Nordic Region: Exceptionalism, Migrant Others and National Identities* (29–41). New York: Routledge.

Pfeifer, Pitseolak (2018). From the credibility gap to capacity building: An Inuit critique of Canadian Arctic Research. *Northern Public Affairs*, 6(1): 29–34.

Rasmussen, Derek (2002). Qallunology: Pedagogy for the Oppressor. *Philosophy of Education*: 85–94.

Rasmussen, Derek (2004). Cease To Do Evil, Then Learn To Do Good: A Pedagogy for the Oppressor. In C. A. Bowers and F. A. Marglin (Eds.), *Rethinking Freire: Globalization and the Environment* (115–130). Mahwah, NJ: Lawrence Erlbaum.

Simpson, Audra (2007). On Ethnographical Refusal: Indigeneity, 'Voice', and Colonial Citizenship. *Junctures*: 67–80.

Smith, Linda T. (2012). *Decolonizing Methodologies: Research and Indigenous Peoples*, (2nd ed.). London/New York: Zed Books.

Thuesen, Søren, Gulløv, Hans C., Seiding, Inge, Toft, Peter A. (2017). Erfaringer, ekspansion og konsolidering 1721–82. In H. C. Gulløv (Ed.), *Grønland – Den arktiske koloni* (46–107). Copenhagen: Gads Forlag.

Tomaselli, Keyan G., Dyll, Lauren, Francis, Michael (2008). "Self" and "other": autoreflexive and indigenous ethnography. In N. K. Denzin, Y. S. Lincoln, and L. T. Smith (Eds.), *Handbook of critical and indigenous methodologies* (347–372). Thousand Oaks, CA: SAGE Publications.

Tuck, Eve (2009). Suspending Damage. *Harvard Educational Review*, 79(3): 409–427.

Tuck, Eve andFine, Margaret (2007). Inner angles: A range of ethical responses to/with Indigenous and decolonizing theories. In N. Denzin and M. Giardina (Eds.), *Ethical futures in qualitative research: Decolonizing the politics of knowledge* (45–168). Walnut Creek, CA: Left Coast Press.

Tuck, Eve, Yang, K. Wayne (2012). Decolonization is not a metaphor. *Decolonization: Indigeneity, Education & Society*, 1(1): 1–40.

Tuck, Eve andYang, K.Wayne (2014). Unbecoming Claims: Pedagogies of Refusal in Qualitative Research. *Qualitative Inquiry*, 20(6): 811–818.

Williamson, Karla J. (2014). Uumasuusivissuaq: Spirit and Indigenous writing. *In Education*, 20(2): 135–146.

Wilson, Shawn (2008). *Research is Ceremony: Indigenous Research Methods*. Halifax/Winnipeg: Fernwood Publishing.

Zavala, Miguel (2013). What do we mean by decolonizing research strategies? Lessons from decolonizing, Indigenous research projects in New Zealand and Latin America. *Decolonization: Indigeneity, Education & Society*, 2(1): 55–71.

5 Industrial development in Nuuk and Sermersooq

Empowerment through action research

Allan Næs Gjerding & Ina Drejer

Introduction

This chapter discusses an industrial development project in Greenland's capital city, Nuuk, and the surrounding municipality. The aim is to illustrate the benefits and challenges of action research as a participatory method in projects where the success depends on involvement of and acceptance by constituent actors who need to break free from reified structures.

In this chapter, participatory research is characterized as action-oriented research aimed at providing results for a specific group of actors, drawing on knowledge from these same actors who become co-researchers in the process. Ideally, participatory research enables actors and gives them ownership over the process (Cornwall and Jewkes, 1995). In real life, however, the participants' involvement in the process is unpredictable over time, and there could be different interpretations, agendas, and means for enacting solutions, which challenge implementation (Ibid.). The following pages present a case of a participatory action research project carried out from April 2016 to September 2017 in collaboration with Kommuneqarfik Sermersooq, the municipality in which Nuuk is located.

The main argument for choosing an action research approach was not only that knowledge possessed by local actors was indispensable for the project, but also that the development of a new, shared understanding was essential for implementing the necessitated actions.

The chapter unfolds the purpose, execution, and outcome of the project, which addresses industrial development challenges and opportunities in Nuuk and Sermersooq. Greenland is a place where traditional and modern ways of life co-exist. For several decades (although to varying degrees), there has been a dichotomy—and associated tension—between Nuuk as the representative of modern city life and the small settlements representing the "authentic Greenland" (Grydehøj, 2014; Sørensen and Forchhammer, 2011). It is no secret that a considerable proportion of Greenland's population is struggling to make a living. While some see the settlements as the root of the problem and centralization/urbanization as the solution, others argue that the real problem is an inadequate level of economic activity and that relocating people to the cities would not make any difference if they are subsequently unable to find employment (Hendriksen, 2014). Following this line of argument, focusing on the promotion of industrial development opportunities in Nuuk could serve as a long-term lever for development for the entire country.

The next section describes in greater detail the challenges that were the point of departure for initiating a project aimed at supporting industrial development in Nuuk and Sermersooq. This is followed by a presentation of the applied method and approach, including a general discussion of action research approaches that provides a background for positioning the specific approach applied in the project at hand. The chapter then proceeds to an overview of the outcome of the project before reflecting on the appropriateness of the applied approach and converting the lessons learned from the project into recommendations on how to design and carry out action research in Greenland.

Background: Challenges and the purpose of the project

In 2015 Kommuneqarfik Sermersooq initiated a project on industrial development in Nuuk and Sermersooq, which was initially spurred by a desire to explore and trigger opportunities for industrial development related to the— at that time ongoing—construction of a new container terminal at Sikuki Nuuk Harbour and the planned expansion of Nuuk Airport. As the project (which was carried out by the authors[1] of this chapter) unfolded, however, the focus gradually broadened to a more general analysis of growth drivers building different development scenarios and ultimately ending up with proposing specific action points for strengthening the industrial development in Nuuk and Sermersooq.[2]

Greenland is currently undergoing substantial change in several dimensions. This includes a massive tendency toward urbanization, where the population of Greenland is increasingly becoming concentrated in the municipality of Sermersooq, particularly in its main city and the nation's capital, Nuuk. By 2018 Nuuk had 18,060 registered inhabitants, which is almost one-third of the total population of Greenland (55,877) and represents a 30% population increase since 2000.

International experience has shown how urbanization can support industrial development and economic growth by increasing the critical mass of actors and associated activities. But while Nuuk might be able to enjoy some concentration advantages, it remains a small, (very) geographically isolated town by international standards.[3] Accordingly, standard agglomeration and urbanization theories do not apply to the case of Greenland.

Sermersooq is the center of industrial activity in Greenland: Firms are on average twice as large as the national average, and two-thirds of the total of nationwide aggregated salaries and shares are paid in Sermersooq.

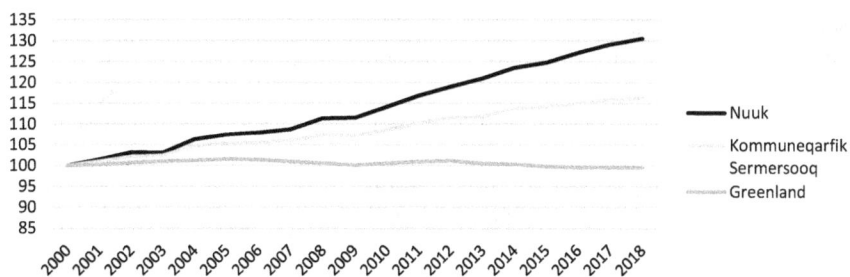

Figure 5.1 Population growth, Nuuk, Kommuneqarfik Sermersooq and Greenland, 2000–2018
Source: Statistics Greenland (index 2000 = 100)

Table 5.1 Population, businesses, and salaries, 2016

	Population	Number of Businesses	Aggregate salaries and shares (DKK, mn)	Average salaries and shares (DKK, '000)
Kommuneqarfik Sermersooq*	22,480 (40.3%)	1,417 (33.6%)	4,272 (65.6%)	3.01
Greenland	55,847	4,212	6,511	1.55

Source: Statistics Greenland
*Percentages express the share of the total for all of Greenland.

Although Nuuk and Sermersooq are doing relatively well compared with Greenland as a whole, the capital city and surrounding municipality do not escape the challenges facing Greenland in general in relation to the ability to generate levels of economic activity capable of supporting the desired level of welfare and wellbeing of the population.[4] The quest for economic growth has qualitative and quantitative dimensions specifically aimed at reducing social exclusion (Kommuneqarfik Sermersooq, 2016).

In addition to the aforementioned limited size and geographical isolation associated with the challenges related to recruitment opportunities, trade conditions, and opportunities for positive synergies and spillovers, the heavy dependence of Greenland's economy on fisheries poses yet another challenge. The dominance of one line of business leaves the economy vulnerable: Since 2000, the proportion of seafood in total Greenlandic exports has, with few exceptions,[5] fluctuated between 85% and 95% of total exports, with no general tendency toward a declining dependence on this sector.

Whereas fisheries are close to being the only source of international trade revenue for Greenland, fishing, together with hunting and agriculture, accounts for only 17% of the principal occupations in Greenland (the corresponding figure for Sermersooq is 8%). There is very little manufacturing in

Sermersooq and Greenland (just above 1% of total employment and 2.5% of industrial valued added), just as the proportion of employment in business services is relatively low (5% in Greenland and 7% in Sermersooq). Hence, 65% of the principal occupation positions in Sermersooq are in either public administration and services (41%), transportation (13%), or wholesale (11%).[6, 7] Accordingly, a major aim of industrial development policy in Nuuk and Sermersooq is to spur economic growth though the development of new business opportunities.

Industrial development is, however, restricted by another well-known challenge, which is present to varying degrees throughout Greenland: access to qualified labor. In 2016 a total of 53% of the labor force in Greenland had no formal education beyond primary school. In addition to being more exposed to unemployment, those without secondary or higher education are also more likely to be excluded from the labor force.[8] Because educational institutions and academic jobs in government institutions and similar organizations are concentrated in Nuuk, the proportion of the labor force with a secondary or higher education is likely to be higher in Nuuk and Sermersooq than elsewhere in Greenland. Nonetheless, although updated statistics on the general level of the qualifications of the labor force are unavailable for Nuuk and Sermersooq, the 2017 quality report from the primary schools in Sermersooq documents that educational attainment is also a challenge here: Among the most recent primary school graduates, 38% were not enrolled in education three months after completing primary school.[9] An analysis carried out by the Sermersooq Business Council confirms that local access to qualified labor is a challenge in Nuuk (Sermersooq Business Council, 2017).

The above-mentioned challenges were the starting point for the project initiated in 2015, which was linked to the preparation and implementation of a Capital Strategy for developing Nuuk as the capital of the Arctic (Kommuneqarfik Sermersooq, 2016). The project also set out to address an additional challenge, however, which affects the capacity to successfully

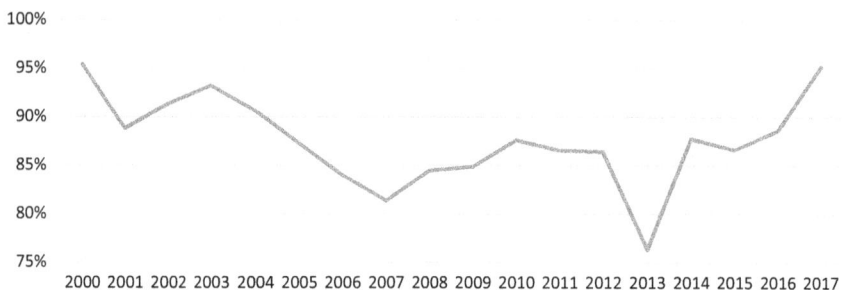

Figure 5.2 Exports of fish, crustaceans, and mollusks in total exports
Source: Statistics Greenland

implement specific actions to strengthen industrial development in Nuuk and Sermersooq. This challenge is related to trust and collaborative culture. Although Greenland's population is small, it is by no means a homogenous country. Grydehøj (2014) characterizes Nuuk itself as an "island city of immense contrasts" (p. 4), with strongly differentiated neighborhoods ranging from modern architecture to broken-down apartment blocks. There is also a stark contrast between Nuuk and the rest of Greenland, where the common perception is that while Nuuk is home to one-third of the national population, it is not the authentic Greenland—in fact, it is barely Greenlandic at all (Sørensen and Forchhammer, 2011; Grydehøj, 2014). These contrasts are reflected in the collaborative culture, as between the government of Greenland, Naalakkersuisut, and Kommuneqarfik Sermersooq, where Naalakkersuisut, in its quest to counterbalance the influence of modern Nuuk, can in some cases be seen as having interests that are at odds with the capital city. This is illustrated by Sermersooq's decision in 2014 to leave Kanukoka, the national association of municipalities in Greenland. Sermersooq subsequently expected to be able to negotiate directly with Naalakkersuisut on national matters—which the Premier of Greenland at that time declined.[10] However, it is not only between different public authorities that collaboration can be lacking: Businesses in Greenland are also reluctant to collaborate (Gjerding et al., 2012).[11]

Accordingly, the project set out to address the following:

1 Barriers for interaction between physical planning, infrastructure, and transportation patterns as a means to support industrial development
2 Competence-related barriers for the industrial development of the Greenlandic business community
3 Barriers for collaboration between businesses, public authorities, and organizations on industrial development issues

These targets reflect the inherent existence of "wicked problems" (Rittel and Webber, 1973), in the sense that the understanding of what constitutes the problems and their possible solution evolve over time as new constituent actors become involved. In order to overcome the wicked problem, the actors must form shared understandings by co-producing situated knowledge (Cunliffe and Scaratti, 2017) resulting in policy changes as the exchange of ideas among constituent actors is translated into concrete action (Karlsen and Larrea, 2014).

Accordingly, the project was organized as an action-oriented research process in which the researchers not only observe but also seek to contribute to addressing the outlined challenges through interactions with core actors. This is articulated in the contract with Kommuneqarfik Sermersooq in terms of a request for proposing specific actions that can alleviate the emphasized short- and long-term challenges.

Approach and method

In the following section, we outline the approach and method of the present study. First, we position our study in the diverse and vivid plurality of approaches that constitutes contemporary action research. This positioning is summarized in a number of criteria that we find useful in assessing the quality of action research. Second, we present how our study unfolded and discuss how the study meets the criteria.

Positioning the study within action research

Action research covers a wide range of approaches and methods (Chandler and Norbert, 2003; Cassell and Johnson, 2006; Reason and Bradbury, 2008; Bradbury, 2015). What is common to the multiplicity of scientific endeavors is the focus on solving problems and achieving change by empowering actors who are key to the knowledge concerning the problems at hand and possible avenues for their resolution. However, this focus is pursued from different paradigmatic stances and with the use of methodological triangulation to such an extent that action research can be described as "not a theory or a method but a strategy for using multiple theories and methods opportunistically" (Greenwood, 2015: 199). This implies that action research is eclectic and that action research-oriented projects are difficult to categorize within specific paradigms, because they are likely to combine different worldviews and practices for uncovering problems, the causes of problems, and how such problems can be addressed.

Despite the eclectic nature of action research, there are limits to the inclusiveness of paradigmatic positions. Action researchers engage in research in the pursuit of an emancipatory agenda. This can be more or less ambitious, ranging from minor projects to improve practice, for example, in health care, education, or business, to projects that strive to deconstruct colonial practices within a post-colonial context. This implies that the researcher becomes an actor who is engaged in a specific social context where research is based on co-producing knowledge with practitioners for the greater purpose of changing the current state of affairs. In effect, action research does not favour paradigmatic points of departure where the role of the researcher is to uncover objective truth at arms-length; instead, it favors approaches of a critical nature, where the emphasis is on interpreting and co-producing practical knowledge and visions (Waterman et al., 2001; Wicks, Reason, and Bradbury, 2008; Bradbury, 2015a; Greenwood, 2015). This interpretivist stance implies that critical realism, critical hermeneutics, social constructivism, and phenomenology are dominant paradigmatic positions within action research, and that within this broad paradigmatic field we find frequent applications of critical theory, feminism, pragmatism, interactionism, and systems and complexity theory (Brydon-Miller, Greenwood, and Maguire, 2003; Greenwood, 2015).

Whichever the approach, the basic tenet of action research is the pursuit of empowerment through dialogue. This idea stems from the seminal work of Lewin (1948) on how to resolve conflicts in organizations, especially in circumstances of organizational change. While dialogue within action research takes place in numerous ways, as might be expected from the proliferation of approaches, there seems to be general agreement that dialogue must be based on knowledge that is situated in the sense that it is embedded in the context in which dialogue takes place. As argued by Cunliffe and Scaratti (2017), the purpose of this is to mobilize the combination of "expertise, tacit and explicit knowledge about our lived contextualized experience that needs to be surfaced and understood" (p. 30). The focus on empowerment implies that the participants in action research are not restrained from airing their views and experiences and that different perspectives within a group of participants are treated respectfully and carry weight in developing shared understandings and visions. Therefore, a centerpiece of action research is to create "safe spaces" wherein "the participants can be confident that their utterances will not be used against them, and that they will not suffer any disadvantages if they express critical or dissenting opinions" (Bergold and Thomas, 2012: 4).

Safe spaces have two dimensions. The first concerns the degree of inclusiveness of actors in the field to which action research is applied (i.e., which and how many actors we include in the co-research process). The second dimension concerns the length of time in which action research is applied (i.e., what is the time horizon of the project in question). The answers to these questions obviously depend on the research objectives and how these objectives are framed by practical and financial circumstances. But both questions must be answered in terms of the practical knowledge that is produced, meaning that the dimensions must be designed so that the production of practical knowledge becomes useful for the participants in the action research project. This will normally require a cyclical process in which the participants have the opportunity to create, reflect upon, and enact co-produced knowledge and insights (Kemmis and McTaggert, 2000; Coughlan and Coghlan, 2002; List, 2006; Koshy, Koshy and Waterman, 2011; Coghlan and Brannick, 2014).

Co-producing situated knowledge in action research cycles means that the action researcher and the action co-researchers develop shared understandings of what must be changed and how to do so. This is a kind of dialogical sensemaking (Cunliffe and Scaratti, 2017) that can be pursued through various forms of dialogical methods. A widespread method within action research cycles is the dialogue conference (Gustavsen, 1992), which originally "evolved as a negation of classical procedures between the labour market parties" (Gustavsen, 2001: 18). This kind of consensus-seeking dialogue represents a pragmatic orientation as opposed to the critical orientation of action research that focusses on how emancipation can be achieved by liberating resistance to existing circumstances (Johansson and Lindhult,

2008). As explained by Johansson and Lindhult (2008), these two orientations have originally been compared with traditions developed in, respectively, the industrialized and so-called Third World,[12] where the industrialized tradition reflected joint collaboration within a generally accepted consensus, while the Third World tradition reflected the dissolution of antagonistic and unequal social relations (Brown and Tandon, 1983). In both cases, the researcher serves as an enabler of change with the participation of co-researchers, albeit in very different settings that have come to be known as, respectively, participatory action research and participatory research (Cassell and Johnson, 2006). While participatory action research employs a pragmatic orientation, participatory research is more inclined to adopt a critical orientation. Common to both orientations is their focus on how to re-freeze reified structures. As explained by Nielsen and Lyhne (2016), reified structures can be "perceived as being frozen institutional patterns that are part of everyday life" (p. 56). Reified structures confine people within a set of beliefs and behaviors that reproduce the circumstances from which action research aims to deliver the co-producers of practical knowledge. The role of the researcher is to contribute to a process by which alternatives to reified structures are surfaced. This renders the researcher a co-creator of the field to be studied and acted upon.

Action research is value driven. In effect, it does not meet the ideals of objective science. Whether this means that action research outcomes can be generalized or not is disputed (Blichfeldt and Andersen, 2006; Gustavsen, Hansson, and Qvale, 2008), but since the criteria for scientific validity is not objectivity that can be tested through replication, the extent to which generalization is possible is less relevant to the assessment of whether we are doing good research. The quality of action research can be assessed in various ways. Drawing on Guba and Lincoln (1994) and Holkup et al. (2004), Fraser (2018) argues in favor of authenticity as a criterion of quality. Authenticity comprises ontological authenticity (sharing, acknowledging and respecting the perspectives of others), educative authenticity (ensuring mechanisms that allow ontologies to change as a function of encounters), catalytic authenticity (research methods encourage action), and tactical authenticity (research methods participate in empowering participants) (Fraser, 2018: 209).

Asking "How do we know when we are doing good research?", Bradbury (2015a) suggests seven criteria that "are the product of a 'collogue' among editors of *Action Research* journal" (p. 9), which we summarize in the seven questions shown in table 5.2:

The unfolding and quality of the present action study

The research reported in this chapter is characterized by a pragmatic orientation to dialogical sensemaking. As will become clear in the following, the action research criteria for doing good action research are clearly discernible,

Table 5.2 Criteria for good action research

Question	Purpose
1. Are the objectives explicitly addressed?	The objectives of the research must be explicitly addressed in order to secure that they inform how research is undertaken
2. Are participative values and the relational component of research included?	Co-researchers must have a formative role in the research process to generate empowerment and authenticity
3. Does the research contribute to developing our knowledge?	Research must contribute to dialogical sensemaking and/or the advance of theoretical knowledge
4. Are processes and methods clearly stated?	Processes and methods are articulated in a way that clearly guides the research process and creates a transparent analysis
5. Does the research provide new ideas that guide action?	Research must contribute to policy change and the unfreezing of reified structures
6. Are the researchers aware of the role they play in the action research process?	The researchers must continuously reflect on how they are co-producers of the field that they study
7. Are the research insights significant in a wider context?	The research must contribute not only with context-specific insights, but also to our general understanding of the studied phenomena

Source: Elaborated from Bradbury (2015a)

while authenticity is pursued to a modest extent. The pragmatist orientation is informed by a critical approach in the sense that the project activities aim at changing current states of affairs. The action research cycle within the project is a single loop; that is, iterations of co-production of practical knowledge do not occur. The project entails a critical utopian element (Nielsen and Nielsen, 2006; Nielsen and Lyhne, 2016) in the sense that it outlines alternative futures and arrives at practical action points. These practical action points are derived from a process inspired by scenario-building (Schwartz, 1996; Mietzner and Reger, 2005) and by the idea of future workshops where the present is criticized, the future is imagined, and avenues to the future are contemplated (Jungk and Müllert, 1987).

The process by which the critical utopian element evolved comprised three phases. Phase 1 was a field study in Nuuk, where the researchers gathered qualitative and quantitative data, had informal talks with civil servants employed in the municipality, explored the city and its facilities in order to gain a first-hand impression of the milieu, and engaged in conversations with 14 key actors in Nuuk's industrial and educational life.[13]

The researchers produced a document on the drivers for growth and potentials and barriers for industrial development (Gjerding and Drejer, 2016a) that was subsequently discussed with representatives from the municipal authorities. Phase 2 was undertaken by the researchers in solitude, who developed scenarios for the future industrial development of Nuuk and Sermersooq Municipality (Gjerding and Drejer, 2016b). The scenarios reflected the knowledge produced in Phase 1 and portrayed alternative future states of affairs. Phase 3 focused on developing action points that could be subsequently implemented. The core of Phase 3 was a workshop on the scenarios and possible avenues for realizing future opportunities where the co-producers of knowledge in Phase 1 were supplemented by other key actors in Nuuk's industrial life, totaling 28 workshop participants (including the researchers). On the background of the workshop, the researchers summarized the workshop insights in specific action points grouped in five focus areas (Gjerding and Drejer, 2017).

Authenticity of action research was pursued in a number of ways. Regarding *ontological authenticity*, the Phase 1 conversations focused on developing shared understandings in the individual conversations. The researchers developed an agenda for the conversations which was adapted to the role that the conversation partner played in the local community. This agenda was not fixed, however, and the conversations were allowed to unfold according to their own dynamics. In effect, as the conversations evolved, the situated knowledge of the conversation partners was allowed to unfold and become part of a co-production of practical knowledge. This process was reiterated in the Phase 3 workshop, where the participants developed shared understandings in groups that were subsequently presented and discussed across groups, in effect creating a plurality of practical knowledge that could subsequently be formalized in actionable propositions. While the researchers' impression was that the arena for creating practical knowledge was a safe space, we cannot be entirely sure of this. As Nuuk is a small community within a rather limited physical space, when airing points of view, actors can take into account that they do not assume an anonymous position, but rather a public one. In effect, the airing of such viewpoints might be subjected to self-imposed censorship. Actually, the researchers experienced indications of interpretations of expressed views being affected by a confirmation bias. This can be explained by the fact that actors belong to social groups and communities of practices that influence their perceptions and thoughts about future avenues of action (Gray and Gabriel, 2018). This challenge is always present in action research (see, for example, Williamson and Prosser, 2002; Moosa, 2013).

The focus on action points and the process by which these points were created was a solid foundation for *catalytic* and *tactical authenticity*. However, the catalytic and tactical authenticity depends on how action points are subsequently adopted and enacted in the local community. Unfortunately, as described below, the continuation of the process was disrupted by local authorities, owing to discontinuity in key staffing, so *educative authenticity* was not really achieved, which weakened the catalytic and tactical authenticity. That participants' involvement over time is unpredictable is a well-known challenge in action research.

The *criteria for good action research* were largely fulfilled in the present project. The *research objectives* were explicitly addressed in several ways and formed the basis for conversations and desk research in Phase 1; were developed in Phase 2 as a result of the Phase 1 co-production of localized knowledge; openly discussed and elaborated upon in the Phase 3 workshop; and finally reframed as actionable propositions in the action point report that concluded Phase 3 (Gjerding and Drejer, 2017). During and between these phases, the objectives were discussed among researchers and municipality staff at meetings and video conferences. *Participative values and the relational component of research* were clearly present, as the Phase 1 conversations, Phase 2 scenarios, and Phase 3 workshop all formed an evolving process of co-producing localized knowledge and images of preferable future actions and states of affairs. The researchers and co-researchers explicitly discussed that the co-researchers had been selected, because they represented both a plurality of perspectives and key positions in the local community with respect to realizing avenues for development and change. The *contribution to knowledge development* gradually evolved in the sense that the Phase 1 activities led to the formulation of scenarios in Phase 2 that subsequently resulted in shared understandings of actionable propositions in the Phase 3 workshop. This process was a kind of dialogical sensemaking that led to the development of localized practical knowledge in the form of action points related to specific and important opportunities and challenges. Throughout the project, *processes and methods* were clearly stated, both in the reports documenting Phases 1 and 2; during the co-research conversations; and in the purpose and structure of the Phase 3 workshop. *New ideas for guiding action* were co-produced during the Phase 3 workshop and formalized in the final document on action points. *Awareness* of the role that the researchers played in the action research process was not only discussed among the researchers, but also made explicit to the co-researchers during conversations and workshop. Similarly, the role of co-researchers as co-producers of practical knowledge and as key to subsequent actions was addressed in the conversations. Finally, the project *insights* appear to be significant in a wider context, as the drivers, opportunities, and barriers identified in Phase 1, the scenarios developed in Phase 2, and the actionable propositions emerging from Phase 3 are important and relevant not only to the Nuuk area, but to the society of Greenland in general. Furthermore, the problems and opportunities in focus in the project also seem similar to the states of affairs in other Arctic areas (Ozkan and Schott, 2013), so even though the practical knowledge produced by the project might not be directly applicable to other Arctic areas, it will at a minimum serve as a source of inspiration for action and change.

Outcome of the Action Research Project

Action research projects with a pragmatist orientation essentially have two outcomes: The first is that situated knowledge within a plurality of co-researchers is translated into shared practical knowledge that can form the basis for actionable propositions. The actionable propositions represent a new shared understanding that has evolved from dialogical sensemaking, where the contextualized experiences of the co-researchers have been shared through explicit exchanges in the researcher–co-researcher interplay. The second is that the actionable propositions are brought to the next step; that is, to stop being propositions and become actions undertaken by co-researchers and other community actors.

Regarding the *first outcome*, the new shared understanding in the present action project evolved gradually over the course of the phases of the project.

In Phase 1, the conversations revealed numerous drivers for growth, together with opportunities and barriers in relation to industrial development. The drivers included:

- infrastructure investments
- improvements to the digital infrastructure
- increasing competition within a centralized economy
- securing access to financing
- continuing urbanization and concentration of the population
- investments in human capital
- macroeconomic stability combined with an increasing openness to international trade and finance

The opportunities and barriers related, respectively, to specific industries (i.e., fisheries and tourism) and to the general dynamics of industrial evolution. The situated knowledge that was shared and developed during the conversations was made explicit and formalized in a document prepared by the researchers containing a large number of observations and viewpoints that had emerged from the conversations and were subsequently distilled into six types of opportunities and seven types of barriers (Gjerding and Drejer, 2016a). These distilled opportunities and barriers reflected a major challenge that appeared in the conversations: that the future social and economic development of the Nuuk area and Greenland in general is highly sensitive to the current clash between the interest to preserve indigenous cultural traits and practices and developing cultural traits and practices associated with a modern European and global life style. This clash essentially reflects a major political and cultural struggle within Greenlandic society on how to develop a future Greenland identity. The conversations also revealed how industrial development is primarily pursued by individual actors, such as major companies, the municipality, and the government, all pursuing their own goals in parallel, despite growing recognition of the need to coordinate activities across the set of actors. This clash reflects how, even though Greenland is largely a planned economy, antagonistic relations flourish between the decision-making actors across various levels and sectors of society.

Phase 2 translated these recognitions into scenarios for further co-research (Gjerding and Drejer, 2016b). The researchers developed these scenarios in a very conventional fashion by juxtaposing opposing forces in two dimensions. The first concerned a tradition-bound vis-à-vis modern industrial development, whereas the second dimension related to the lack of and need for the coordination of activities that was described in terms of a disjunctive vis-à-vis a conjunctive state of affairs. Four scenarios for industrial development were described, namely tradition-bound disjunction, modern disjunction, tradition-bound conjunction, and modern conjunction (see figure 5.3).

In Phase 3, these scenarios were presented and discussed at length in the workshop, and the group work resulted in a large set of observations and propositions focused on two major themes:

- How can we improve the cooperation between the key actors in Sermersooq?

  a Who are the key actors?
  b What are the challenges and barriers in relation to cooperation?
  c What are the potentials and motivation of cooperation?
  d Ideas for concrete activities and projects?

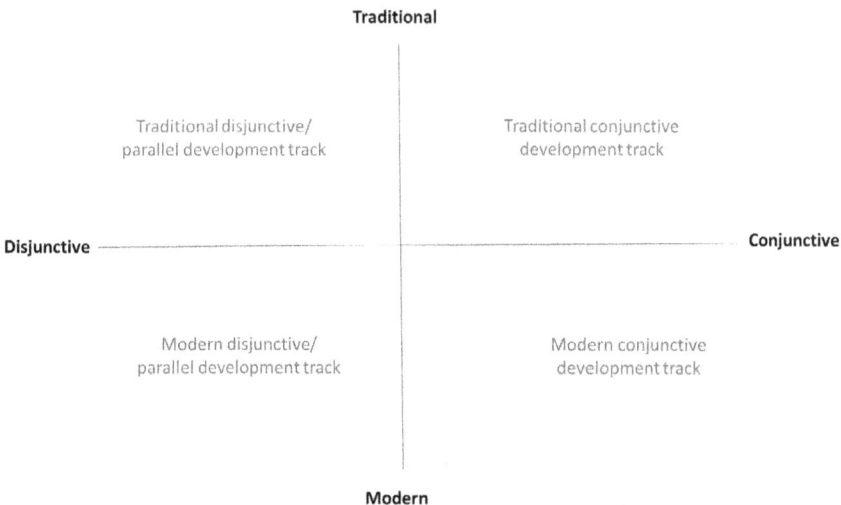

Figure 5.3 Scenarios for industrial development in Nuuk and Sermersooq
Source: Gjerding and Drejer (2016b)

- Are the tradition-bound and the modern prerequisites or opposites?

  a   When do we experience that the tradition-bound and the modern collide?
  b   How do we stimulate development based on the tradition-bound?
  c   Is there a need to "sacrifice the sacred?"
  d   Ideas for concrete activities and projects?

The workshop evolved along the lines of a future workshop in which the present was assessed, the future imagined, and avenues to the future were suggested. Subsequently, the researchers condensed the abundant observations and ideas into 19 action points covering five major themes:

- Rejuvenating practices and decision-making in the public sector (six action points)
- Sharing practical knowledge and providing transparency via positive storytelling (three action points)
- Stimulating openness to the international society, identity work within the Greenland society, and cooperation among key actors (three action points)
- Developing human resources (three action points)
- Stimulating industrial development at the level of the firm (four action points)

Regarding the *second outcome*, the actionable propositions were not brought to the next step (i.e., becoming real-life actions). The number of action points was deliberately kept to a minimum in order to make actions manageable and provide immediate guidance for future action. The actor for initiating the next step was Sermersooq Municipality, where one co-researcher in particular served as the key driver. However, because he subsequently left the municipality to seek an employment opportunity abroad, there was a discontinuation of the involvement, which delayed the initiation of the actions. Instead, the action points entered further policy development at the municipal authority level,[14] the extent and format of which is currently unknown to the authors. This meant that while reified structures were clearly identified and challenged during the project, the re-freeze and subsequent dissolution of reified structures are still in the making.

Conclusion: Reflections on the appropriateness of the applied method and recommendations for action research in Greenland

The method applied in this action research study is considered appropriate in the sense that it largely fulfilled the criteria for good action research. It also enabled the conditions for action research authenticity and resulted in numerous actionable propositions that could enter policy schemes and initiatives. However, while the project did challenge the logic of existing reified structures, the timeframe available for the project and the policy-making capacity of the local community were not sufficient to be able to secure real changes to the practice and dynamics of industrial development. Addressing the complexity of the wicked problems would have required a considerably longer timeframe to generate concrete policy changes. Consequently, a few recommendations and qualifications are in order. These recommendations and qualifications relate not only to the project in question, but are generally applicable to similar projects characterized by the traditional–modern life dichotomy.

The development of scenarios was extremely useful to this project. The interviews undertaken in Phase 1 revealed important political and practical tensions regarding the traditional–modern balance and the ability (and probably also willingness) of local actors to undertake concerted efforts. Using these tensions as opposites—a classical method in scenario-building—allowed the researchers to visualize future states of affairs and possible routes to these stages, which energized the Phase 3 workshop discussions. This energizing effect resulted from the actors' focus moving from the present state of affairs to possible future states of affairs that could become the subject for the joint development of ideas and actionable propositions. In the present case, this meant that ongoing disputes were put aside in favor of the conceptualization of imaginable actions, business propositions, and solutions to obstacles for social and economic development.

Obviously, it is of paramount importance that key authority actors assure that the human resources necessary for action are present, both in terms of the ability to process actionable propositions into policy schemes and the ability to act accordingly. If this requirement depends on one actor, the likelihood of a standstill of the further application of practical knowledge is high. In effect, actions and change cannot depend on one key actor; rather, they must be undertaken by a plurality of consorted actors. In the present case, this implies a need to undertake at least one additional cycle of co-research in which the researchers are commissioned to engage in the realization of action points, taking co-research to the next level. If this is not the case, the project risks becoming a consultancy project rather than a case of participatory action research. This all implies that the time horizon of the project must be adjusted to the opportunities for action. Sufficient time must be allocated to follow up on actionable propositions, where the researchers can become actors in the realization of ideas, especially in cases where there is a lack of human resources in the local community.

Notes

1 The project was commissioned by the municipality of Sermersooq at the Center for Logistik og Samarbejde (Center for Logistics and Cooperation) in Denmark, which is a company owned by Port of Aalborg. Prior to the commission, the center

had been actively pursuing project assignments in Greenland as part of a cooperation agreement between Greenland and Aalborg Municipality, which owns the
Port of Aalborg. As part of the close cooperation on projects between the Center
for Logistics and Cooperation, and Aalborg University, the Department of Business
and Management at Aalborg University, where the authors are members of staff,
was asked to undertake the project.

2 Whereas the new container terminal at Sikuki Nuuk Harbour was inaugurated in
September 2017, the expansion of Nuuk Airport with an extended runway has
been repeatedly postponed. At the time of writing, construction had yet to begin.
However, a national plan was approved in November 2018 by the Greenland Par-
liament, Inatsisartut. This involves the inclusion of Denmark as co-owner of
Greenland's two planned additional international airports (Nuuk and Ilulissat).

3 Furthermore, urbanization possibly also has a downside in Nuuk in terms of the
high degree of social vulnerability of socioeconomically disadvantaged newcomers
to the capital.

4 The pursuit of increased industrial development and associated economic growth
may also be related to a growing national desire to make Greenland more—ultimately completely—independent of Denmark.

5 In 2013, the proportion of fish, crustaceans, and mollusks in total Greenlandic
exports dropped to a historically low level of 76%. There is no single explanation
for this, but exports of "ships, aircraft, and drilling rigs and production platforms"
were exceptionally high in that year, and the price index for cod (the main export
fish item) dropped by more than 20 points compared with 2012. Source: Statistics
Greenland.

6 The corresponding percentages for Greenland as a whole are 39%, 9%, and 11%.

7 This section is based on data drawn from Statistics Greenland. Information on
value added is based on accounting statistics, which are available for limited or
private limited companies and foreign branches only (669 out of the 4,212 businesses registered in Greenland in 2016).

8 Source: Statistics Greenland.

9 https://sermersooq.gl/da/borger/skolernes-kvalitetsrapport-2017/#cat=274, accessed
18 August 2018.

10 Later, other municipalities also left Kanukoka, and the association was dissolved
in 2018.
    Sources: https://knr.gl/da/nyheder/kanukoka-har-sidste-dag-31-juli; http://sermitsia
q.ag/node/171667;   https://naalakkersuisut.gl//da/Naalakkersuisut/Nyheder/2015/02/
240215-Naalak-bredt-samarbejde; accessed 2 September 2018.

11 Although some argue that resistance to collaboration is less pronounced among
businesses in Nuuk than elsewhere in Greenland (Gjerding et al., 2012).

12 Greenland is an interesting case in relation to this dichotomy, because it is
characterized by a mix of properties associated with advanced and developing
economies alike.

13 These actors were identified by the municipality as key actors in Nuuk's industrial
and educational life. However, several of the actors did not perceive themselves as
key actors in the sense of feeling included in the ongoing policymaking process.

14 According to follow-up information obtained by the authors.

References

Bergold, J. and Thomas, S. (2012). Participatory research methods: A methodological
approach in motion. *Forum Qualitative Sozialforschung/Forum: Qualitative Social
Research*, 13(1): Art.30.

Blichfeldt, B. S. and Andersen, J. R. (2006). Creating a wider audience for action research: Learning from case-study research. *Journal of Research Practice*, 2(1): Art. D2.

Bradbury, H. (Ed.) (2015). *The SAGE Handbook of Action Research* (3rd ed.). Dorchester: SAGE Publications.

Bradbury, H. (2015a). Introduction: How to Situate and Define Action Research. In H. Bradbury (Ed.), *The SAGE Handbook of Action Research* (3rd ed.) (1–12). Dorchester: SAGE Publications.

Brown, L. D. and Tandon, R. (1983). Ideology and political economy in inquiry: Action research and participatory research. *The Journal of Applied Behavioral Science*, 19(3): 277–294.

Broydon-Miller, M., Greenwood, D., and Maguire, P. (2003). Why action research? *Action Research*, 1(1): 9–28.

Cassell, C. and Johnson, P. (2006). Action Research: Explaining the Diversity. *Human Relations*, 59(6): 714–783.

Chandler, D. and Torbert, B. (2003). Transforming inquiry and action. *Action Research*, 1(2): 133–152.

Coghlan, D. and Brannick, T. (2014). *Doing Action Research in Your Own Organization*. Dorchester: SAGE Publications.

Cornwall, A.Jewkes, R. (1995). What is participatory research? *Social Science and Medicine*, 41(12): 1667–1676.

Coughlan, P. and Coghlan, D. (2002). Action research for operations management. *International Journal of Operational & Production Management*, 22(2): 220–240.

Cunliffe, A. and Scaratti, G. (2017). Embedding impact in engaged research: Developing socially useful knowledge through dialogical sensemaking. *British Journal of Management*, 28(1): 29–44.

Fraser, S. L. (2018). What stories to tell? A trilogy of methods used for knowledge exchange in a community-based participatory research project. *Action Research*, 16 (2): 207–222.

Gjerding, A. N. and Drejer, I. (2016a). *Centrale aktørers opfattelse af vækstdrivere, muligheder og barrierer for erhvervsudvikling i Nuuk og Sermersooq*. Aalborg: Department of Business and Management, Aalborg University. http://vbn.aau.dk/files/261780454/Sermersooq_procesdokument_Fase1_endelig_version.pdf.

Gjerding, A. N. and Drejer, I. (2016b). *Udviklingsscenarier: Udviklingsveje for den fremtidige erhvervsudvikling i Nuuk og Sermersooq*. Aalborg: Department of Business and Management, Aalborg University. http://vbn.aau.dk/files/261780478/Scenarier_for_erhvervsudvikling_i_Nuuk.pdf.

Gjerding, A. N. and Drejer, I. (2017). *Hvordan styrkes erhvervsudviklingen i Nuuk og Sermersooq?*Aalborg: Department of Business and Management, Aalborg University. http://vbn.aau.dk/files/262149469/Hvordan_styrkes_erhvervsudviklingen_i_Nuuk_og_Sermersooq_afrapportering_fra_workshop_om_barrierer_og_muligheder_Fase3.pdf.

Gjerding, A. N., Hansen, P., and Lysholm, A. P. (2012). *Behovet for erhvervsøkonomiske kompetencer i det grønlandske erhvervsliv* (report). Aalborg: Aalborg University. http://vbn.aau.dk/files/70331768/Behovet_for_erhvervs_konomiske_kompetencer_i_Gr_nland_WP_2_12.pdf.

Gray, D. E. and Gabriel, Y. (2018). A community of practice or a working psychological group? Group dynamics in core and peripheral community participation. *Management Learning*, 49(4): 395–412.

Greenwood, D. J. (2015). An analysis of the theory/concept entries in the *Sage Encyclopedia of Action Research*: What we can learn about action research in general from the encyclopedia. *Action Research*, 13(2): 198–213.

Grydehøj, A. (2014). Constructing a centre on the periphery: Urbanization and urban design in the island city of Nuuk, Greenland. *Island Studies Journal*, 9(2): 205–222.

Guba, E. G. and Lincoln, Y. S. (1994). Competing paradigms in qualitative research. In N. K. Denzin and Y. S. Lincoln (Eds.), *Handbook of Qualitative Research* (105–117). Thousand Oaks, CA: SAGE Publications.

Gustavsen, B. (1992). *Dialogue and development*. Assen: Van Gorcum.

Gustavssen, B. (2001). Theory and practice: The mediating discourse. In P. Reason and H. Bradbury (Eds.), *Handbook of Action Research* (17–26). Trowbridge: SAGE Publications.

Gustavsen, B., Hansson, A. and Qvale, T. U. (2008). Action research and the challenge of scope. In P. Reason and H. Bradbury (Eds), *The SAGE handbook of Action Research* (2nd ed.) (63–76). Trowbridge: SAGE Publications.

Hendriksen, K. (2014). Grønlands økonomi i et bosætningsperspektiv. *AG Grønlandsposten*.

Holkup, P. A., Tripp-Reimer, T., Salois, E. M., and Weinert, C. (2004). Community-based participatory research: An approach to intervention research with a Native American community. *Advances in Nursing Science*, 27(3): 162–175.

Johansson, A. W. and Lindhult, E. (2008). Emancipation or workability? *Action Research*, 6(1): 95–115.

Jungk, R. and Müllert, N. (1987). *Future Workshops: How to Create Desirable Futures*. London: Institute for Social Inventions.

Karlsen, J. and Larrea, M. (2014). The contribution of action research to policy learning: The case of Gipuzkoa Sarean. *International Journal of Action Research*, 10 (2): 129–155.

Kommuneqarfik Sermersooq. (2016). Nuuk – Arctic capital: Capital strategy for Nuuk. Report, June.

Koshy, E., Koshy, V., and Waterman, H. (2011). *Action Research in Healthcare*. London: SAGE Publications.

Lewin, K. (1948). *Resolving Social Conflicts*. New York: Harper & Row.

List, D. (2006). Action research cycles for multiple future perspectives. *Futures*, 38(6): 673–684.

Mietzner, D. and Reger, G. (2005). Advantages and disadvantages of scenario approaches for strategic foresight. *International Journal of Technology Intelligence and Planning*, 1(2): 220–239.

Moosa, D. (2013). Challenges to anonymity and representation in educational qualitative research in a small community: A reflection on my research journey. *Compare*, 43(4): 483–495.

Nielsen, H. and Lyhne, I. (2016). Adding action to interview: Conceptualizing an interview approach inspired by action research elements. *Action Research*, 14(1): 54–71.

Nielsen, K. A. and Nielsen, B. S. (2006). Methodologies in action research: Action research and critical theory. In K. A. Nielsen and L. Svensson (Eds), *Action Research and Interactive Research: Beyond Practice and Theory* (63–87). Maastricht: Shaker Publishing.

Ozkan, U. R. and Schott, S. (2013). Sustainable development and capabilities for the polar region. *Social Indicators Research*, 114(3): 1259–1283.

Reason, P. and Bradbury, H. (Eds) (2008). *The SAGE Handbook of Action Research* (2nd ed.). Trowbridge: SAGE Publications.

Rittel, H. W. J. and Webber, M. M. (1973). Dilemmas in a general theory of planning. *Policy Science*, 4: 155–169.

Schwartz, P. (1996). *The Art of the Long View.* New York: Doubleday.

Sermersooq Business Council. (2017). *Rekruttering og kompetencer for virksomheder i Nuuk,* http://da.business.gl/fileadmin/user_upload/documents/SBC_KompentenceRap_071217_DK_web.pdf, accessed 18 August 2018.

Sørensen, B. W. and Forchhammer, S. (2011). Byen og grønlænderen. In O. Høiris and O. Marquardt (Eds), *Fra vild til verdensborger: Grønlandsk identitet fra kolonitiden til nutidens globalitet* (559–596). Aarhus and Copenhagen: Aarhus University Press.

Waterman, H., Tillen, D., Dickson, R., and de Koning, K. (2001). Action research: A systematic review and guide for assessment. *Health Technology Assessment*, 5(23).

Wicks, P. G., Reason, P., and Bradbury, H. (2008). Living inquiry: Personal, political and philosophical groundings for action research practice. In P. Reason and H. Bradbury (Eds), *The SAGE Handbook of Action Research* (2nd ed.) (15–30). Trowbridge: SAGE Publications.

Williamson, G. R. and Prosser, S. (2013). Action research: Politics, ethics and participation. *Journal of Advanced Nursing*, 40(5): 587–593.

6 Collaboration to secure relevance and quality in a study of EIA practise in extractive industries in the Arctic

Sanne Vammen Larsen & Anne Merrild Hansen

Introduction to the research project and its subject

In the past decade, increased global demand for hydrocarbons has contributed to a growing international interest in hydrocarbon exploration and production in the Arctic. In Greenland in the period 2002–14, licenses for hydrocarbon activities have been subject to bidding rounds every second year (Naalakkersuisut, 2014, Østhagen, 2012). In 2017 and 2018 licensing rounds were held again for areas in Baffin Bay and Davis Strait (Naalakkersuisut, 2018a). While falling oil prices have led to some resignation from companies operating in the Arctic, exploration licenses remained active in the waters northwest and northeast of Greenland in 2016 during the project presented here and remained so in late 2019 (Naalakkersuisut, 2016; Naalakkersuisut, 2018b). Hydrocarbon exploration has been taking place in Greenland since the 1970s and hardly represents a new phenomenon. In pursuit of increased independence from the former colonial power, Denmark, and the implementation of a system of self-rule government, however, in 2009 the Greenland Government, Naalakkersuisut, decided to take over the administration of mineral resources, including hydrocarbons, which was previously administered by the Danish authorities. A new Greenlandic Mineral Resources Act was developed (later amended in July 2014), with the subsequent implementation of new guidelines in 2015 for environmental impact assessment (EIA) scoping and the environmental impact assessment of seismic activities.

EIAs are intended to ensure informed and transparent decision-making processes as well as more sustainable activities. These aims are pursued by identifying, assessing, and mitigating the environmental impact of proposed activities, publishing the results, and carrying out public participation initiatives as part of the process. EIAs are mandated by legislation in many countries worldwide, and although the legislation is not the same, it is based on the same basic template and covers many of the same activities (Glasson, Therivel, and Chadwick, 2012; Senécal et al., 1999). This creates the potential to seek inspiration in how legislation, procedures, and processes are designed in other jurisdictions with similar conditions. In Greenland, there are currently legal requirements for companies to conduct a series of impact assessments related to various phases of hydrocarbon development (Naalakkersuisut, 2014). The assessments include Strategic Environmental Assessment (SEA), Environmental Impact Assessment (EIA), and Social Impact Assessment (SIA). Together with the approval system and guidelines, impact assessments serve to ensure that hydrocarbon activities are developed in a sustainable manner, where environmental and social considerations are incorporated early in relation to new projects and therefore inform decision-making processes. The assessments and where in the phases of the development of a hydrocarbon project they take place can be seen in Figure 6.1.

Offshore hydrocarbon activities can cause major damage to the environment if environmental protection plans are not proactively incorporated in industrial processes, as they can potentially lead to transboundary pollution (e.g., oil spills). It is therefore important that legislation is at the highest international level and in accordance with so-called best practice. The Arctic is particularly sensitive, from an environmental perspective, and Greenland has received massive international attention from environmental organizations in relation to hydrocarbon exploration. This has been seen in Greenpeace and WWF campaigns, and reports from Lloyds and Ernst & Young have affirmed the importance of the strong environmental regulation of oil exploration activities (Ernst & Young, 2013; Lloyds, 2012). They explicitly emphasize the international importance of monitoring and developing environmental legislation in accordance with best practice as part of the necessary environmental protection efforts.

Based on the above, a study was initiated in 2014 to investigate how the Greenlandic regulation of the EIA of hydrocarbon activities could be developed through an international comparison of EIA legislation for hydrocarbon activities in Greenland, Denmark, Norway, Canada, and the US state of Alaska.

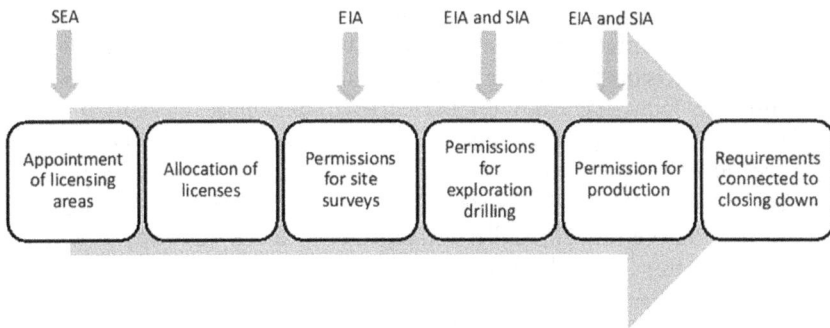

Figure 6.1 The impact assessments required by Greenlandic legislation for the various phases of a hydrocarbon project

The collaborative approach of the research project

In this section, the design and methods of the study are presented, focusing first on the comparative case study and the collaborative approach and subsequently on the specific methods for data collection.

The study was a comparative study covering Greenland, Denmark, Norway, Alaska, and Canada. Norway, Alaska, and Canada were chosen based on their lengthy experiences with hydrocarbon activities in the Arctic Region. Denmark is included in the study based on the close administrative and political collaboration with Greenland, providing a potential for the transfer of regulation.

The study involved the following stages:

1   Gathering and initial screening of documents
2   Scoping of study
3   Analysis of documents
4   Supplementary data collection
5   Comparative analysis
6   Interpretation of findings

Research conducted in collaboration with local communities is generally acknowledged as good practice and applied by many (particularly social) research practitioners throughout the Arctic (Grimwood et al., 2012). The practice of what is often referred to as "community based participatory research" covers a range of action-oriented intentions, participation, and collaborative partnerships (Stewart and Draper, 2009). The experience of researchers and community members with collaboration is that the application of collaborative methods in research not only substitutes paternalistic and condescending practices of research *on* communities but also supports and activates the capacity inherent in local knowledge systems (Battiste, 2008). Benchmarking the Greenland legislation on impact assessment related to hydrocarbon development in Greenland and choosing a collaborative approach seemed relevant for us to allow for a degree of local control over the research processes and outcomes, but also to foster trust through transparent, reciprocal, and interactive relationships, and, last but probably most importantly, to support the community ownership of results in order to secure informed priorities and decision-making through knowledge generation.

In order to focus the study on the current challenges relevant to Greenlandic administrative practice, a collaborative element was included in two stages of the study (referring to the stages previously presented):

• Stage 2: Before the data collection started (to focus the study).
• Stage 6: After data collection (to interpret the findings).

In both of these stages, a workshop was arranged in Greenland's capital, Nuuk, with invited stakeholders. Table 6.1 presents more details from the workshops.

Table 6.1 Overview of the two workshops held as part of the study

Activity	Participants	Date
Workshop Stage 2	Naalakkersuisoq (Minister) of Mineral Resources Secretary to the Minister Head of the Agency for Mineral Resources Administrative officer from the Agency for Mineral Resources Chairman of Transparency Greenland Representative from KNAPK, Association of Fishers and Hunters in Greenland Representative from Greenland Oil Industry Association Two representatives from Shell Greenland Representative from Nunaoil (National Oil company) Rector from University of Greenland Researcher from University of Greenland Researcher from Greenland Institute of Natural Resources	February 24, 2015
Workshop Stage 6	Head of the Agency for Mineral Resources Two administrative officers from the Agency for Mineral Resources Head of secretariat from Transparency Greenland Representative from Association of Fishers and Hunters in Greenland Representative from Nunaoil (National Oil company) Two researchers from Greenland Institute of Natural Resources Researcher from the Arctic Centre at Aarhus University, Denmark	October 21, 2015

The invited and represented stakeholders included the main authorities, non-governmental organizations (NGO), industries, and research institutions involved in the planning and EIA processes concerning hydrocarbon activities in Greenland. The content and methodology of the workshops are described in detail in the following sections.

Workshop 1 (Stage 2)

In the first workshop, the participants were asked, based on their professional knowledge of the Greenlandic context, which themes concerning the EIA of hydrocarbon activities they found most relevant. The workshop involved 13 participants and was held at Hotel Hans Egede in Nuuk on February 24, 2015. The program included an introduction, group discussions, presentations, and feed-back in plenary.

The workshop was introduced by the authors of this chapter, who were the two researchers engaged in the project. After explaining the purpose of the study, the overall frame and delimitation, the methods and timeline, and finally the preliminary results from Stage 1 were presented. The preliminary results consisted of an overview of the hydrocarbon development process and the application of impact assessment in this process. Three groups were then formed and encouraged to suggest themes they found interesting and relevant to investigate in the study. The groups presented the output of their discussions to each other in plenary and received feedback from the other participants. They were then asked to prioritize the themes according to importance. There was general consensus regarding the challenges and themes that the participants found relevant to investigate further.

Overall, the participants emphasized a primary interest for the analysis to focus on ongoing exploration activities and decision-making processes in Greenland rather than on potential activities. On this basis, decision was made to focus the study on environmental assessments conducted in relation to exploration drilling, as this was the current stage of hydrocarbon development in Greenland. The study therefore did not cover the later stages in development, such as EIA or SIA for appraisal drilling, construction, production, closure, or monitoring.

The participants highlighted the following themes as being of importance: public participation, EIA process and reporting, legislative design and governance practice, baseline data, knowledge production, impartiality, and legitimacy. After prioritizing the themes, three stood out as the most important to address in the study: 1) public participation in EIA processes, including the mandate of the citizen and decision-making competence; 2) databases and standardization, including local knowledge; 3) assessment methods, including the use of alternatives and evaluation of significance as well as the role of the citizen in this regard.

Further information (in Danish only) about the second workshop can be found in (Hansen and Larsen, 2015a).

Workshop 2 (Stage 6)

The second workshop was also held at Hotel Hans Egede in Nuuk and took place in the afternoon on October 21, 2015. Nine people participated in the workshop. The participants from the first workshop had been invited again, but we received several cancellations. Some institutions appointed other representatives, and in practice this meant that only three people attended both workshops. This second workshop involved stakeholders in interpreting the findings, in the form of differences and similarities in EIA regulations between the five countries. The aim was for them to point out areas where they found that Greenland could potentially learn from the regulation in the other countries.

The workshop began with an introduction, including a summary of the purpose and process of the study. It was described how input from the first workshop had influenced the shape and content of the study, and then the findings from the study were presented. The presentation was structured according to the three themes prioritized in the first workshop, including findings from all of the countries included in the study. The results were continuously discussed in plenary. After the presentation and initial discussion, two groups were formed. The groups were asked based on the results of the document study to identify which results they assessed—on the basis of their professional knowledge of the Greenlandic context—should be part of the discussion of the future development of EIA of hydrocarbon activities in Greenland. The conclusions from the groups were presented and discussed in plenary.

Overall, there were three different outputs from the second workshop: firstly, four focus areas for improvement of the Greenlandic EIA legislation and practice were identified; secondly, knowledge gaps were identified; and, thirdly, findings/conclusion were nuanced and corrected by the participants.

Further information (in Danish only) about the second workshop can be found in a report by Hansen and Larsen (2015b).

Main findings

The workshop participants identified three main areas they found to be of importance to the development of the Greenlandic regulation of the EIA of hydrocarbon activities:

1   Public participation in EIA processes
2   Databases and standardization
3   Assessment methods

This is a result unto itself, as it points out what the stakeholders find to be the weak points in the current system. The study was therefore designed to focus on these three overall themes.

The main findings of the study are divided into two parts: a range of points where the EIA demands differ (which can be used to discuss the pros, cons, and lessons for Greenland) and a list of four detailed focal points of special relevance for development in Greenland.

General points for discussion

The analysis produced numerous issues for the discussion of how to strengthen the Greenlandic EIA system:

• Should environmental assessment related to decisions on licensing areas be stronger? How would this influence the need for EIAs in the lifecycle of oil projects?

- Should public participation in the process of identifying new licensing areas be stronger?
- Should social impacts be integrated in environmental assessments or be continuously subject to individual assessments?
- Should the focus of public participation in different phases of the EIA be specified?
- Should the authorities be made more responsible for public participation (potentially with financial support from operating companies)?
- Should the financial support system for the public related to EIAs be improved by developing more specific guidelines?
- Should a report be required in relation to EIAs describing how the public has influenced decision-making and the process?
- Should a clear guide for filing complaints be made for EIAs?
- Should authorities be able to further define the format of data in baseline studies and EIAs to allow for data to be shared more easily in common databases and potentially made available to the public?
- Should there be requirements regarding the use of alternatives and the assessment of significance in EIAs?

Focal points

Based on the above results, four focal points were chosen as those to which the local stakeholders ascribed particular importance. The first focal point of special relevance in Greenland is providing support for the public to find and finance external aid and counselling during the process, such as to help to read and understand the often long, technical EIA reports. This point is inspired by the Canadian participant funding scheme. The second focal point concerns access to complaints, where systems and approaches in Denmark and Alaska inspired a focus on a more transparent and standardized access to complaints, making it easier and less complicated for the public. Inspired by the Norwegian and Canadian processes, there was also a call for earlier participation in the project lifecycle. This should secure public input at the strategic level, such as when it is decided which areas should be opened up for hydrocarbon activities. An SEA is currently carried out during this phase of hydrocarbon projects (see Figure 6.1) but does not include open public participation. Finally, a point arose that did not come from any of the studied countries, but rather from general discussions with participants, namely the need for more interactive methods for public participation than those usually mentioned in legislation in all countries; the written hearing plus public meetings.

As can be seen from both the general points for discussion and the focal points, the discussions and focus of the stakeholders in the second workshop focused mainly on public participation as the crucial issue. For more information (in Danish only) about the results from the study, see Hansen and Larsen (2016).

Discussion of the applied approach

The experience of the researchers involved was that the organization of the workshops using work in smaller groups was productive. The opportunity to talk in smaller groups facilitated the exchange between the participants and meant that a civil tone was kept between stakeholders who are otherwise often on opposing sides in decision-making processes. The smaller groups meant that each participant had more room and time to speak, which seemed to contribute to a richer exchange of experiences, building of capacity, and better knowledge of the challenges and viewpoints of the other stakeholders.

This leads to an important point in how to get people to engage and participate, which is not always an easy task. In this case, the research team was happy with the group that was put together, which included stakeholders who do not often participate in such activities. The research team did not have many challenges finding participants. Several factors could be relevant here, our assessment being that a key element is what the participants can gain from participating—"What's in it for us?" In this case, the research project, topic, and the workshops had the attention of key stakeholders, not least the minister. This creates attention and renders participation attractive to other stakeholders because it provides a realistic opportunity to gain influence. Thus, stakeholders might prioritize participation when they can see that other key stakeholders will be present. As discussed above, the opportunity to build capacity and exchange knowledge can also be an asset that promotes participation; an asset that also depends on the participation of key stakeholders who are respected for their knowledge and experience in the field.

Another issue is that of how to assemble the group of stakeholders to invite questions, such as how many participants and which ones? When bringing representatives from NGOs, authorities, science, and industry together, how does one strike the right mix or balance so that the various interests are represented equally? In our case, these questions were tackled mainly through the prior knowledge of the local stakeholders and networks. One important consideration was to avoid having too few participants to be able to form groups while at the same time avoiding so many that it would not be possible to create a friendly and intimate atmosphere, where the participants felt it was safe to share and discuss their input. One way of going about choosing who to invite in such a case, where we worked with stakeholders, is to carry out a structured stakeholder analysis to provide an overview.

Many of the issues in this section are relevant for this case—because it involved professional stakeholders, the approach, discussion, and results could have been very different if it had instead involved ordinary citizens. In a bigger picture, the stakeholders involved in our study are all to some extent part of the educational elite of Greenland, and they all work with these issues professionally. Thus, the question of whether they represent the broad population can arise. Some of them are obviously meant to represent others in their functions, such as the representative from Kalaallit Nunaanni Aalisartut Piniartullu Kattuffiat which is the interest group for hunters and fishermen. Still, the study might have looked differently and produced different results if we had instead worked directly with "ordinary" citizens from different backgrounds—or possibly with the politicians as a subgroup. Several aspects are in play here. One is that it might have been more time-consuming to involve a representative segment of the Greenlandic population. Another is what the purpose and target groups of the study are—whether it is aimed at improving the usefulness of EIAs for ordinary people or going more into the subtleties aimed more at the professionals working with EIAs from different approaches. For the purpose of this study, the approach of working with professionals was chosen. In our experience, it is of the highest importance for participatory research to reflect over who should be involved and what that means for the study, its results, and ultimate their application.

Although the research team did not encounter problems assembling stakeholders for collaboration, we did experience changes in participants between the two workshops. This was a mix of different circumstances, such as:

- Who was in Nuuk at the time of the workshop, as many of the stakeholders also operate in other parts of Greenland and the world
- Changes in staffing in the relevant organizations between the two workshops
- General availability at the specific time (since the research team travelled for the workshop, the date and time was not very flexible)

Especially the first two issues appear more predominant in the Arctic than in other places we have worked. The changes in participants between the two workshops meant an emphasis on the need for a thorough recap at the beginning of the second workshop and that we needed to think about giving the participants time to connect and reconnect at the beginning, since they did not all know each other (and us) from the first workshop.

As stated in section 2, the first part of the collaborative process was partly intended to help the researchers to focus the study on issues of relevance to the stakeholders (the study was focused on the three stakeholder-prioritized issues). At the same time, so many specific issues for analysis were suggested at the workshop that the researchers later found it necessary to choose from the many possible issues, together with issues of importance from a research perspective. For example, some of the issues suggested at the workshop in relation to public participation were the timing of the EIA process, who is involved in the public participation, who is responsible for the participation, how the participation process is announced, and what is considered good public participation in the five countries. This emphasizes the need for clarity about the extent of the participation and how much power the stakeholders have to decide the details of the study through their input. With just a single two-hour workshop, it might only be feasible for the stakeholders to provide input to the overall direction of the study and suggestions for specific points of analysis, leaving decisions on the detailed scope to the researchers. This should be communicated to the participants if relevant, making clear how their input will be used and processed.

An important topic we often discuss with research colleagues who are also applying collaborative methods in research in Greenland and other Arctic countries is the language barrier, which often occurs between researchers and the locals engaged in the projects. Many researchers active in Greenland do not speak the native tongue, Kalaallisut. Moreover, it can be difficult to find interpreters, and paying for their travels and their pay, among other things, adds additional costs to projects. This means that researchers often conduct interviews and engage with their field of inquiry more generally in the language of the researcher. Obviously, this means that some individuals and stakeholder groups are excluded or poorly represented. Even when stakeholder groups are represented by persons who speak the language of the researcher (typically Danish or English) as their second or third language, researchers have experienced cultural issues and barriers affecting communication. This can potentially complicate and even distort results. This primarily becomes a challenge when attempting to relate to the general public.

The research team itself has assessed that the collaborative approach produced valuable local knowledge about the EIA regulation and its development, as well as helping to anchor the project and its results in the local context, meaning that some of the results were immediately applicable. For example, the Agency for Mineral Resources started work on new standard guidelines for access to complaints on the basis of the results of the project. In hindsight, other collaborative methods could have been used, such as interviews or workshops with smaller, more defined groups. This might have enabled the research team to probe deeper into the details and backgrounds of the input and viewpoints of the actors, and they might also have been even more open than they were when placed in a larger group of peers. However, individually or in smaller groups, we would probably not have gotten the constructive discussions and common viewpoints and recommendations, as they might have been more set on their own interests and uninterested in compromising. Moreover, the actors would have missed the opportunity to meet and exchange experiences and knowledge. The research team is therefore satisfied with the methods used for the purpose of this research project.

Conclusion and recommendations

The preliminary reflections about the use of collaborative methods in the project are that they helped to secure the relevance of the analysis and final recommendations, as well as an anchoring of the results that meant that they were used. Moreover, on the participant side, there was an exchange of knowledge and experience, together with a chance to influence important stakeholders. The following lessons can be drawn from the discussions above:

- Consider how much influence the participants can and should have on the study and be clear about the boundaries to match expectations
- Working in smaller groups in breakout sessions works well
- Make it clear to the participants what they might gain from their collaboration
- When recruiting participants, be open about who else is invited and/or have agreed to participate
- Consider carefully who to involve based on the purpose and target groups of the research project and be aware of what this means for the interpretation and use of the results
- Be aware that continuity in participants in a process over time can be a challenge in many Arctic contexts. Try to mitigate the problems, for example by providing a good recap and taking time to get (re-)acquainted

References

Battiste, M. (2008). Research ethics for protecting Indigenous knowledge and heritage: Institutional and researcher responsibilities. In N. K. Denzin, Y. S. Lincoln, and L. T. Smith (Eds), *Handbook of Critical and Indigenous Methodologies* (497–509). Thousand Oaks, CA: SAGE Publications.

Ernst & Young. (2014). *Exploring Arctic Oil and Gas: Challenges to Arctic Resource Recovery.* Retrieved December 1, 2015 from: www.ourenergypolicy.org/wp-content/uploads/2013/09/Arctic_oil_and_gas.pdf.

Glasson, J., Therivel, R., and Chadwick, A. (2012). *Introduction to EIA* (3rd ed.). London: Routledge.

Grimwood, B. S. R., Doubleday, N. C., Ljubicic, G. J., Donaldson, S. G., and Blangy, S. (2012). *Engaged Acclimatization: Towards Responsible Community-based Participatory Research in Nunavut. The Canadian Geographer/Le Géographe Canadien,* 56(2): 211–230. doi:10.1111/j.1541-0064.2012.00416.x.

Hansen, A. M. and Larsen, S. V. (2015a). *Benchmarking af miljøvurderingslovgivning for olieaktiviteter i Grønland. Bilagsrapport 1—Opsamling fra Workshop* 1. Aalborg: Danish Centre for Environmental Assessment.

Hansen, A. M. and Larsen, S. V. (2015b). *Benchmarking af miljøvurderingslovgivning for olieaktiviteter i Grønland. Bilagsrapport 2—Opsamling fra Workshop* 2. Aalborg: Danish Centre for Environmental Assessment.

Hansen, A. M. and Larsen, S. V. (2016). *Miljøvurdering af offshore kulbrinte aktiviteter: En benchmarkingundersøgelse af krav til VVM for efterforskningsaktiviteter i Grønland, Norge, Canada, Danmark og Alaska.* Aalborg: Danish Centre for Environmental Assessment.

Lloyd's. (2012). Arctic Opening: Opportunity and Risk in the High North. Retrieved December 1, 2015 from: www.lloyds.com/news-and-risk-insight/risk-reports/library/natural-environment/arctic-report-2012.

Naalakkersuisut. (2014). Our mineral resources—Creating prosperity for Greenland: Greenland's oil and mineral strategy 2014–2018 quick read version. Retrieved June 2016 from: http://naalakkersuisut.gl/~/media/Nanoq/Files/Publications/Raastof/ENG/Olie%20og%20min eralstrategi%20ENG.pdf.

Naalakkersuisut. (2016). List of mineral and petroleum licenses in Greenland (June 2016). Retrieved June 2016 from: www.govmin.gl/images/list_of_licences_20160602.pdf.

Naalakkersuisut. (2018a). Overview of hydrocarbon licensing rounds in Greenland, September 2018. Retrieved September 2018 from: www.govmin.gl/en/licensing/app lications/hydrocarbon-licensing-rounds.

Naalakkersuisut. (2018b). List of mineral and petroleum licenses in Greenland, September 2018. Retrieved September 2018 from: www.govmin.gl/images/Documents/Current_Licences_and_Activities/List_of_Licences_03-09-2018.pdf.

Østhagen, A. (2012). Dimensions of Oil and Gas Development in Greenland. Washington, DC: The Arctic Institute—Center for Circumpolar Security Studies. Retrieved from: www.govmin.gl/images/Documents/Current_Licences_and_Activ ities/List_of_Licences_03-09-2018.pdf.

Senécal, P., Goldsmith, B., Conover, S., Sadler, B., and Brown, K. (1999). *Principles of Environmental Impact Assessment Best Practice*. Fargo, ND: International Association for Impact Assessment.

Stewart, E. J. and Draper, D. (2009). Reporting back research findings: A case study of community-based tourism research in northern Canada. *Journal of Ecotourism*, 8 (2): 128–143.

7 Critical proximity in Arctic research

Reflections from the Arctic Winter Games 2016[1]

Carina Ren & Robert C. Thomsen

Introduction

This chapter is written on the basis of empirical data collected from the 2016 Arctic Winter Games (AWG) held in Nuuk, Greenland (Arctic Winter Games, 2016), to illustrate and discuss different research approaches within the field of community- and identity-building, as well as the opportunities and downsides of their interlinked research positionalities. Based on experiences gained from our engagement in and fieldwork during the AWG 2016, we invoke the concept of "critical proximity" (Birkbak et al., 2015; Latour, 2005: 253) as a fruitful way to think about research and related activities, not as external to social life but as its co-produced outcome. As we demonstrate, such an approach differs from how collective, social identities are typically researched in the Arctic context and offers, as we show, new and locally anchored ways to understand identity positions and identity-building.

We argue that participatory and collaborative approaches create the possibility for more engaged and, thus, comprehensive research, but also discuss its contestedness, as close connections challenge our understanding of critical distance as paramount to conducting "proper" research at a distance (Corby, 2017; Hayward and Cassell, 2018). As we hope to illustrate, critical proximity allows for new ways of seeing and understanding the interventionist capacities of research as it is conducted together with (as opposed to at a distance from) the actors who make up our field of study.

In order to develop this argument, we first offer a review of literature on the topic, which provides the reader with a brief overview of the methodologies deployed in research into community- and identity-building in the Arctic. We propose three dominant approaches in this field of research. The first, which we term *"desk-observational"* is characterized by distant critique, whereas the second, *"present-inclusive,"* is based on the presence of the researcher(s) in the community, as well as by some inclusion of research subjects. We term the third (and still largely absent) research approach *"collaborative co-creational."* Next, we introduce the research project, a brief presentation of the AWG 2016, and our aim of exploring such endeavors in relation to identity and societal value creation.

We use two vignettes to illustrate the processes and outcomes of critical proximity research and to describe how research "interfered" with the field and how, in turn, the field came to shape many aspects of the research. Drawing on these insights, we discuss the pros and cons of researchers growing entangled with the field and sketch out ways forward for more critical proximity in Arctic research.

Exploring researcher–field relations in Arctic research

In the following section, we offer a simple typology and selected examples of how researchers have positioned themselves and their studies in relation to the general field of collective identity-building in Arctic research.[2] Recent decades have seen a rapidly growing number of research projects and publications on Arctic themes from virtually all academic disciplines. Consequently, various coordinating units, organizations, and committees have deemed it necessary to set out some basic, ethics-based guidelines for proper research conduct in the region.

At the time of writing, for example, the American Interagency Arctic Research Policy Committee (IARPC) is seeking comments on its newly revised *Principles for Conducting Research in the Arctic.* Although these principles "are directed at [U.S.] federally-funded researchers," it is also suggested that "they may be useful to academic, state, local, and tribal researchers in the Arctic" (International Arctic Social Sciences Association, 2018: para. 2). Here, and on its website, IARPC lists five core principles: "Be accountable; Establish effective two-way communication; Respect local culture and knowledge; Build and sustain relationships; and Pursue responsible environmental stewardship" (Interagency Arctic Research Policy Committee, 2018: para. 2). The items that talk about two-way communication, respect for local culture and knowledge, and sustainable relationships between researchers and communities seem to point to participatory research as core to principled Arctic research.

Other national and transnational organizations, including the Danish Forum for Arctic Research, have developed similar guidelines. Under the heading "Relationship between researcher and local population," its guidelines address the importance of inclusion and dialogue (Forum for Arktisforskning, 2018: 2). Although such principles seem laudable, it must be acknowledged that various kinds of research inherently suggest various degrees of the exercise of such behavior: A researcher's methodological approach is largely insisted upon by the academic traditions and disciplines within which she/he is trained and contributes. Anthropologists, political scientists, epidemiologists and climatologists have been trained within—and thus traditionally adhere to—different epistemological tenets and, consequently, carry different methodological toolboxes.

From the humanities and social sciences, contributions to Arctic research are made from a broad range of disciplines, including anthropology, ethnology, sociology, history, political science, law, and linguistics, and even within these, methodological approaches differ. In a sub-field of research that can broadly be termed "community and collective identity research,"[3] much scholarly work has been carried out as what one might term *"desk-observational"* research (see, for example, Dahl, 2010; Fondahl, 1999; Gad, 2017; Hicks and White, 2015; Loukacheva, 2007; Thisted, 2011, 2017; Thomsen and Vester, 2016). This kind of research has contributed greatly to building academic knowledge about historical developments and contexts, structures, discourses, and practices, and in that sense, the methodological approach has proved itself to be valuable. For certain other kinds of research purposes, however, this approach falls short.

In some disciplines, the tradition has therefore been for closer involvement with the community "under scrutiny," allowing for a larger degree of research data to be harvested with the collaboration and active participation of local informants, interviewees, etc. The principles proposed by the IARPC and the Forum for Arktisforskning (Forum for Arctic Research) would seem to suggest that scholars from different disciplines increasingly observe that kind of practice; that there be a general movement toward greater involvement of the communities and the people who constitute the object of study into what can be described *"present-inclusive"* research.

Within the for-this-purpose broadly defined field of community and identity research, a substantial amount of valuable research has been carried out in that form (see, for example, Bjørst, 2017; Hansen, 2017; Heine, 2013; Rygaard, 2010; Sejersen, 2015; Shadian, 2014; Skille, 2013). The academic methodological tradition and the physical challenges of transport to/from and within the Arctic continue to form obstacles to this kind of proximity-based research for many, while the spread of effective online communication provides new vistas for it.

In other cases, scholars venture into an approach to Arctic community and identity research termed here as *"collaborative co-creational."* This approach is characterized by a high level of integration between researcher(s) and actors in the field as concerns, research questions, and methods of the project are collaboratively negotiated and developed throughout the research process. All the way into its outcomes, it is an iterative and highly enmeshed process (see, for example, Arnfjord, 2014; Arnfjord and Andersen, 2014; Berliner et al., 2012; Gesink et al., 2010; Pedersen and Rygaard, 2003a, 2003b). Wiita (2006) identifies a movement in social science research in the circumpolar north which, in her opinion:

> is becoming more collaborative—with local residents and other scientists. This is a positive and necessary change in the culture of research. The research arena is striving for scientists to partner with local residents rather than work in isolation from them in their communities and on their lands. (p. 4)

As Alia explains:

> there has been a revolution in research methodology in the past two decades. In place of the old pattern in which researchers descended on Arctic communities and left with artifacts and information, today's physical and social scientific research is conducted in close collaboration with Arctic residents and sometimes has considerable impact on Arctic policy and community projects and programs.
>
> (Alia, 2005: 11559, quoted in Wiita, 2006)

In addition to what is arguably an ethical requirement within certain fields to perform more collaborative research, such an approach also holds obvious benefits toward crafting socially sound, relevant, and response-able research. In this form, the integrated researcher–field relationship becomes advantageous in at least two interrelated ways. Firstly, the researcher gains access to knowledge, perspectives and issues of which they might not otherwise have become aware. Secondly, by way of the general knowledge and theoretical prisms provided by the researcher, individuals in the field who engage in the project will be able, in the process, to qualify and relate their personal experience to the historical processes and structures of society as a whole. In other words, the "sociological imaginations" (Mills, 1959) of both the researcher and the field collaborator are activated and eventually result in the expansion of individual and shared understanding. While important inroads have been made, this form of participatory research within social and cultural identity studies remains comparatively rare.

The above account suggests that we can understand the researcher–field relationship through a continuum between desk-observational and colla-borative co-creational (illustrated in Figure 7.1).[4] As indicated above, our intention here is not to promote one relationship over others or to designate value to either end (or indeed the middle) of this continuum. In one case, critical methodological considerations will suggest a research design characterized primarily by archival, statistical, or other desk-based work; in another, a design consisting mainly of interviews or close collaboration on the research question and data collection is called for. Indeed, larger research projects are likely to benefit from a combination of more than one approach. The model indicates, however, the density and reciprocity of the researcher–field relationship (increasing from left to right).

Desk-observational   Present-inclusive   Collaborative co-creational

The following draws on our experiences with and reflections regarding the study of identity "in the making" during the AWG, when they took place in Nuuk in March 2016. While the research design of the project spans the entire continuum, the examples presented in this chapter serve as

Desk-observational   Present-inclusive   Collaborative co-creational

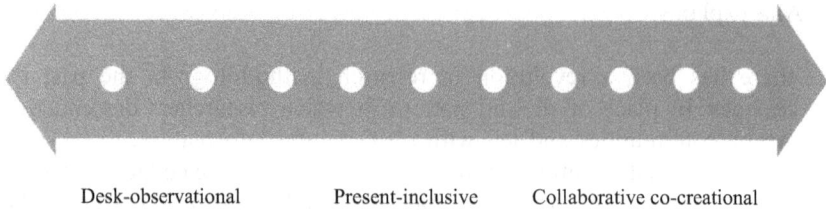

Figure 7.1 The researcher–field continuum

illustrations of collaborative co-creational methods, their benefits, and challenges. First, however, comes an introduction to the concept of "critical proximity," which not only informs our work theoretically but also our methodological approach.

Critical proximity

The current chapter and the research underpinning it are informed by a three-fold interest: ontologically, in how the social is produced also through the encounter between researcher and field; epistemologically, in how this field can become knowable to us through a specific co-creational framing of this encounter; and methodologically, in how research co-creation can lead to the building of knowledge that is able to problematize concerns already present (ed) in the field by its actors.

Our understanding of research positions and positionality in relation to our field is informed by the work of Bruno Latour (2005a) in his warning against what he sees as critical research from a distance. Introducing the concept of "critical proximity," Latour reminds us of how "being critical" does not necessarily (only) imply a researcher critiquing a given phenomenon from an outside and distant position. While researching from a distance might be, as argued above, a logical or at times necessary precondition in Arctic research, it puts us at risk of ordering the world in forceful strokes, which, in *Reassembling the Social*, Latour (2005b) has termed the "panoramas" of critical research. While analytical distance might have its strengths, it also risks blinding us to the task of paying close empirical attention to local concerns and issues.

According to Latour, our research must defy the temptation to critique from a distance and to hastily resort to simple explanatory, often reductionist, frameworks. Instead, we should start with detailed descriptions, which are careful investigations of "what there is in the world" (Law, 2009: 4). Critical proximity to the field blurs traditional divides between research subjects and objects, and it challenges the ideal of objectivity and of "seeing everything from nowhere" (Haraway, 1988). As we shall see, proximity—as opposed to distance—creates further ambiguity. Rather than seeing this as a drawback, however, we understand it as a productive condition, as it also fosters "response-able" knowledge (Haraway, 1997, 2016), in the sense that it enables responses to expressed, relevant exigencies. According to Greenhough (2011), this is the case, as "it signals points of attachment, inherent instabilities, fractures, cracks in the "state of things," which remain open to academic intervention" (p. 136). Lastly, critical proximity grants the field that we acknowledge its own rights and abilities to problematize broad and generic claims about their doings; that is, to bring to the table their version of "what is there in the world." As we shall see in the vignettes, this helps us to abstain from prematurely determining what a given phenomenon is *really* about.

Methodology and case description

The critical proximity concept also shapes much of our methodology in appreciation of how activities carried out by researchers are fundamentally interfering with the world (Law, 2004a) and that research is a performative endeavor, which partakes in world-making. As argued by Vikkelsø (2007), "description is never a passive rendering of the world but a chance for informants to enact certain versions of the world" (p. 306). Thus, critical proximity research does not entail being *more* implicated in world-making through research than other, more "distant" researchers. It does allow us, however, to be implicated with our eyes and ears open, and in so doing to create research that can be more robust and more relevant, as well as more prone toward the multiple ways in which things come together. Considering our concern with the co-creation of research, the question then becomes how we as researchers enable and stage this together with the actors in our research field.

The AWG held in Nuuk in 2016 offers an evident case to exemplify and discuss this. As the largest event of its sort ever held in Greenland and as "more than a sports event" (Arctic Winter Games, 2016), as heralded on the organization's Facebook page, the sports and cultural event was not only an opportunity for circumpolar youth to come together in friendly competition and cultural exchange. It was also an occasion for the Greenland community to offer new stories about what Greenland is and new ways to work together toward this aim. The event thus relates to identity issues in several ways. First, in more "traditional" terms through individual and collective Arctic identity-building amongst participants, and, second, as a more interventionist experimentation with possible future Greenlandic identity through different public, business, and community activities related to the event (Thomsen, et al., 2018).

In studying this event, we could have resorted to a methodology that would have allowed us to make simple judgments—or "panoramas," as Latour proposes—of the event, either as automatically activating identities or, as is often the criticism levelled at large events, simply a waste of money (see Petersen and Ren, 2015). And indeed, the local media reproduced claims that the AWG 2016 would be "a waste of money" (Sermitsiaq, 2012) or "not only too expensive for *Kommuneqarfik Sermersooq* (the municipality) [but] too expensive for the whole country" (Sermitsiaq, 2012). Claims were also made about AWG 2016 being "an elitist project" (informal conversation from fieldwork). However, seeking to exert critical proximity, we were curious also to explore what seeing AWG 2016 as "more than a sports event" would entail, specifically in relation to identity and value creation, our two research fields. This aligns with the advice offered by Birkbak et al. (2015), according to whom "the critically proximate researcher brings forward and supplements the problematization of given issues already happening in everyday and professional practices" (p. 270). Turning attention to what is already grappled with and problematized locally was therefore a major concern.

In order to explore this "at proximity," different methods were deployed, ranging from interviews, workshops, and focus groups, as well as participant observations and "deep hanging out" (Geertz, 1998: 69) at different sites during the event. In the year leading up to the event, Ren also followed the event-planning process closely, which she was allowed to do in exchange for facilitating value workshops for AWG stakeholders and the delivery of a final independent evaluation. While some research situations, such as the workshops with AWG stakeholders in the preparatory stages of the event, were concerned with exploring value creation, others, including the focus groups with Arctic AWG participants, centered on identity. Others, such as participant observation and "deep hanging out" during the event and collecting press and social media content, were relevant to both foci of the study and provided for a broader understanding of how the event tied into everyday activities and stakeholder concerns. For instance: How the event was used to work with corporate social responsibility in sponsoring companies; how Visit Greenland used it as a way to work with influencers; how it was used in sports organizations or at a municipal level to develop a stronger volunteer culture; and how it was used in schools to highlight the benefits of sports.

Vignette 1: How research methodology interfered with the field—and how the field struck back

One of the declared purposes of the Arctic Winter Games is to promote friendships and understanding across Arctic communities—to "bring ... our Circumpolar World closer together" (Arctic Winter Games, 2018). Hence, a subsection of the research project concerned itself with pan-Arctic identity-building among athletes and cultural participants at the AWG. Basic research questions focused on the extent to which such community-building took place and which aspects of the games were (un)conducive to it.

In terms of the continuum introduced above, the approach of this part of the project was "present-inclusive:" much time was spent coordinating focus group interviews with the organizers and *chefs de mission*, designing a semi-structured conversation guide.[5] The methodological basis for the conversation was a thoroughly considered, theoretically supported research design, including considerations about potential practical and ethical challenges, as well as measures to counter such challenges. Altogether, all seemed set to procure much useful empirical data through efficient focus group conversations.

However, unexpected delays owing to extreme weather and consequent rescheduling raised the first practical challenges to the meticulously laid scheme. Ad hoc measures included frantic, last-minute, close collaboration with local organizers and team representatives who struggled together to make the conversations materialize. From a "discreet," minimally disturbing academic event organized by the researchers on the periphery of the games, the conversations thus became more interfering and "noisy" than intended. The orderly process envisaged from an office at Aalborg University became disordered and hectic to the extent that the head researcher at one point (literally) stumbled into a room full of young participants representing various contingents (Alaska, Nunavut, Northwest Territories, Nunavik, Sápmi). These had obviously been herded there on short notice, pulled away from practice, competition, and rest to act as focus group for a research project to which several of them had not been introduced. Owing to a tight time schedule, the room—situated in the local high school—remained organized as a classic classroom: "pupils" in rows, "teacher" in front, rather than the conversation-promoting set-up originally planned.

This was hardly the ideal situation for focus group work prescribed by methodological theory. Indeed, for a while, the atmosphere was anything but conducive to open, trust-based dialogue. Getting participants to engage in conversation proved to be demanding.

Apart from the practical dis-organization of the event, the post-hoc analysis of the situation suggests at least three major challenges to the immediately free-flowing, informative conversation expected: a) The fact that those present belonged to different age groups and different cultural groups (as opposed to the "neatly" organized groups stipulated by the research design) made for a challenge of trans-cultural communication for which the researcher was unprepared and had to address on the spot. b) As a precondition for this conversation to take place had become last-minute "filling up" with basically anybody available, not all of the participants had been prepared for the event. Some of the *chefs de mission* did not appear to take the task seriously and had neglected to introduce their athletes to the project in advance. This meant that there was much confusion about its purpose, which had to be addressed during the conversation. c) In this contingent environment, the researcher's conversation guide proved to be a barrier to fruitful dialogue. Even if setting out a semi-structured conversation (Ennis and Chen, 2012), it insisted on a formality and orderliness that only served to increase the tension of the "teacher–pupil" relationship; to this age group, the researcher–teacher distinction is likely unclear, both presumably considered to carry some sort of intellectual authority. Also, importantly, the issues with which the guide proved to be of little relevance to the young athletes and cultural participants.

The participants seemed to relate to the conversation prompts more out of courtesy than enthusiasm. Thus, initially, they politely tried to respond in a manner that might be expected of them rather than sharing any valuable personal experience. In a subtle way, they in fact refused to be categorized; to accept the premises of predefined identity positions. It was first half-way through the event—after the conversation guide had been laid aside and the researcher decided to "go with the flow" and engage in whichever subject was brought up by the group and offer their own non-academic contributions—that a proper conversation can be said to have evolved, which involved all of the participants.

Cross-cultural dialogue and identity-building indeed proved to be important and valuable to the group, but the composition of identity structures was much more intricate and incoherent than had been expected based on the existing literature. While making no claim here that the agility and open-mindedness of the researcher turned a largely un-productive focus group session into a completely successful one, the unintended "chaos" certainly produced some advantageous effects. One fruitful consequence of "unplanned" critical proximity was that the researcher was forced to take seriously the alternative narratives of "what matters" and adapt to this unexpected intervention in the process of interaction with actors in the identity field. In terms of the researchers' relationship with the field, this part of the empirical work was effectively shifted from a position of "present-inclusive" toward a more "collaborative co-creational" practice.

Negative implications can be argued to have been lacking methodological stringency due to the eventual unawareness of direction and purpose of the conversation on behalf of both participants and researcher. The casual engagement of the researcher and the ensuing blurring of professional–informal lines can also be argued to constitute potential data contamination. The counterargument, however, would suggest that had the entire focus group experience followed the ready-made, "ideal" design, much valuable knowledge would never have been (co-)produced.

Hence, the methodological disruption and ensuing discarding of hauled-along identity boxes allowed insights into otherwise-unknown identity realities. This experience thus suggests the importance of critical proximity in identity research, and it was made abundantly clear in the process that the active involvement of the field in the formulation of core research questions is essential. The argument would be that otherwise-invisible identity positions and practices emerge through such non-academic interventions. Moreover, they provide opportunities for response-able researchers to delve into "what matters" to different groups within the field—and consequently allow them to rethink not just methodology and theoretical assumptions suggested by existing studies and literature but also basic research foci and positionality.

Vignette 2: Exploring 'more than an event'—letting the field do the walking

Over the two years of planning, the event ended up costing over US$10 million divided between Sermersooq Municipality, the Government of Greenland, and the business community. Considering these daunting costs for a one-week event, it would be easy—and perhaps not entirely inappropriate—to criticize it as an untimely expenditure at a time with strained public budgets. A national newspaper wrote an editorial asking: "Can we spend the money better?" (Sermitsiaq, 2012). It would also be easy to label it as a pet project of the Nuuk elite, as many did in comments on social media. Or simply to call it out as an economic flop, pointing out how values were not materializing in the right place, not distributed fairly, or entirely missing.

Issues and discussions of value such as these are reflected in event research and are central to the study of events. In the events literature, events are most commonly explored as economic or sociocultural phenomena. Economic value is typically calculated using quantitative methods, whereas socio-cultural value is explored using more qualitative, at times ethnographic, approaches. Despite the obvious differences in each of these approaches, the critique they generate often centers around whether values are distributed properly between event owners and communities. Thus, despite opposing methodological and philosophical points of departure, both approaches operate in a terrain where values are already "out there" to be counted or criticized for being wrong, missing, or unfairly distributed (Law, 2004b; Ren, Petersen, and Dredge, 2015).

But what if we understood event value as the effects of specific and situated valuing practices and devices in which research also plays a part? This would entail a particular positionality of the researcher challenging the conventional imperative of maintaining a distance to one's object of study. The question then becomes how we can know events and their values in ways, which abstain from distant critique.

Suggestions of what some of these things could be was already hinted at two years before the event kicked off, as AWG 2016 General Manager Maliina Abelsen commented on the newly decided event strategy on Facebook. To her, "AWG is not merely about that week, where everything is launched. It is about those two years [leading up to the event], where we as a community will improve and expand our competences, brand Greenland, collaborate in new ways, and strengthen the areas of sport and culture in general" (Arctic Winter Games, 2016). Thus, already from the very beginning, when the event secretariat was launched by Sermersooq Municipality in the nation's capital to plan and manage the event, the values and outcomes of the event were a concern.

As mentioned, Ren had been working for some time with the event secretariat and other stakeholders and was asked to provide an independent evaluation of the event. While this "bargain" and the invitation to "step into the engine room" of the event planning impeded research detachment, following the argument of critical proximity, it also allowed her to follow closely the recurrent and emergent concerns and discussions on societal values and outcomes. For instance, it became clear through workshops exploring value creation with stakeholders, such as non-governmental organizations, municipal and government officials, and sponsors, that the event was also about other things: collaboration, building trust, and experimenting with work across "silos" and public and private actors. Along the way, it also became clear how the secretariat's efforts to attract 1,500 volunteers were also about service and language upskilling for the tourism industry. This was also connected to the development of a new IT-system to administer volunteers and linked to conversations on how to activate these volunteers in the future to relieve a hard-pressed welfare system.

These are just a few of the event "overflows" (Ren and Mahadevan, 2018) that the critical proximity research position enabled the exploration of—and arguably also itself created. These examples show how critical proximity allows researchers to use local problematization and local concerns of value creation as points of departure for investigation. They show that event values are not simply "out there" (Petersen and Ren, 2015); rather, they emerge—or disappear—through ongoing valuation practices and valuation devices, such as evaluation tools, media, public concern, IT systems, and much more. As argued above, critical proximity grants the field that we study its own rights and abilities to problematize broad and generic claims about their doings together with room to craft other stories of what they are "busy doing."

Conclusion

This chapter has explored and discussed different research–field relations and introduced collaborative research positions based on the concept of critical proximity. As a "critique of critique" and a plea against grand claims about "what there is in the world," critical proximity abstains from too quickly defining what a given phenomenon (e.g., Arctic identity or event valuation at the AWG 2016) is *really* about; instead, it grants the field its own rights and abilities to problematize already-happening issues (Birkbak et al., 2016). In other words, it urges us to initiate and pursue our inquiries within local concerns.

As examples of what such inquiries might look like, the vignettes display our attempts at taking a locally voiced claim seriously that AWG 2016 was "more than an event" and of engaging in the work around "the more" in the field in our research. *Methodologically*, the vignettes show how we did so by putting ourselves (in one case inadvertently) in the middle of things. As described, conducting research at proximity can unsettle traditional research positions as, in the first case, authoritative and "in control," and, in the second, as neutral or value-free. Instead, critical proximity allowed us into a much less foreseeable and more blurred field.

Finding ourselves in this field allowed us to epistemologically expand what we can know by challenging the usually distant analysis that concludes that the AWG was either a "catalyst of identity" or, within the economic sphere, a waste of money. Instead, we encountered spaces where expected categories, distinctions, and values were either missing or messy, as in the case of Arctic identity, or under construction and negotiation, as in the case of the event values. Instead, we were confronted with other and different issues that challenged what the event was about. Active participants were not particularly concerned with Arctic identity-creation; the issue of values and outcomes proved to be about something completely exterior to the event itself: about building competences and strengthening skills, branding Greenland, or rehearsing public–private collaborations.

Besides proposing how a new research positionality and new empirical insights can be generated from a collaborative approach, the vignettes also allow us to draw an "ontological lesson" about how knowledge and realities are co-created through research (Ren et al., 2017). While the ambiguous or entangled outcomes of our research could be addressed as caused by a lack of methodological stringency or distance (see Law and Singleton, 2005), we propose another interpretation. We do not see the messy and entangled knowledge outcomes of research at critical proximity as results of methodological failure or an ethical flaw; rather, they are ontological preconditions, which research must address in productive and response-able ways.

Notes

1 The authors wish to extend our thanks to Steven Arnfjord for the critical reading of an early draft and for directing our attention to omissions and relevant literature in adjacent fields of research.
2 When we use the term "field" here, we suggest that how this field is "populated"— that is, who the relevant actors are to a given research project or, reversely, which actors will find the research of interest to them—cannot be finally determined in advance. Whether we deal with communities, participants, stakeholders, or others is defined in dialogue in the research process. We are also well aware that our review is not exhaustive and therefore cannot be entirely representative of all research carried out with a view to collective identity-building. The purpose is merely to illustrate how the approaches of different studies differ regarding the researcher–field relationship.
3 The sweeping definition of a "community and collective identity research" sub-field can be argued to violate otherwise important distinctions between different types of social, collective identity research (into national, cultural, social, regional, community constructions, etc.). For the specific purpose of this chapter to provide a broad illustration of different approaches, however, it serves an immediate purpose.
4 Others have considered continua and models to illustrate the nature of the researcher–field relationship (see, for example, Arnstein, 1969: 217; and International Association for Public Participation, 2019). To illustrate our own points, however, we have seen fit to develop one for this purpose.
5 In acknowledgement of the age (14–17 years) and inexperience of the participants, the focus group session was described in all correspondence as a "conversation" rather than an "interview" or a "discussion."

References
Alia, V. (2005). Office of Polar programs, National Science Foundation. In M. Nuttall (Ed.), *Encyclopedia of the Arctic* (1st ed., Vol. 1: 11558–11559). New York: Routledge.

Arctic Winter Games. (2016). Arctic Winter Games Nuuk 2016(Facebook page), August 26. Retrieved from www.facebook.com/awg2016/?fref=ts.

Arctic Winter Games. (2018). *Role and value of the Arctic Winter Games.* Retrieved from www.arcticwintergames.org/Role_Purpose_Values.html.

Arnfjord, S. (2014). *Deltagende aktionsforskning med socialrådgivere: Empowerment af Grønlands oversete velfærdsprofession* (PhD dissertation). Nuuk: Institut for Syge- pleje og Sundhedsvidenskab, Ilisimatusarfik/University of Greenland.

Arnfjord, S. and Andersen, J. (2014). Socialt arbejde og aktionsforskning i Grønland. *Dansk Sociologi*, 4(25): 78–93.

Arnstein, S. R. (1969). A ladder of citizen participation. *Journal of American Institute of Planners*, 35(4): 216–224.

Berliner, P., Larsen, L. N., and Soberón, E. C. (2012). Case study: Promoting com- munity resilience with local values—Greenland's Paamiut Asasara. In M. Ungar (Ed.), *The Social Ecology of Resilience: A Handbook of Theory and Practice* (387– 399). New York: Springer.

Birkbak, A., Petersen, M. K., and Elgaard Jensen, T. (2015). Critical proximity as a methodological move in techno-anthropology. *Techné: Research in Philosophy and Technology*, 19(2): 266–290.

Bjørst, L. R. (2017). Arctic resource dilemmas: Tolerance talk and the mining of Greenland's uranium. In R. C. Thomsen and L. R. Bjørst (Eds), *Heritage and Change in the Arctic: Resources for the Present, and the Future* (159–175). Aalborg: Aalborg University Press.

Corby, J. (2017). Critical distance. *Journal for Cultural Research*, 21(4): 293–294.

Dahl, J. (2010). Identity, urbanization and political demography in Greenland. *Acta Borealia*, 27(2): 125–140.

Ennis, C. D. and Chen, S. (2012). Interviews and focus groups. In K. Armour and D. Macdonald (Eds), *Research Methods in Physical Education and Youth Sport* (217– 236). New York: Routledge.

Fondahl, G. (1999). Autonomous regions and Indigenous rights in transition in Northern Russia. In H. Petersen and B. Poppel (Eds.), *Dependency, Autonomy, Sustainability in the Arctic* (55–63). Aldershot: Ashgate.

Forum for Arktisforskning. (2018). *Anbefalinger for god forskningspraksis i Arktis.*

Gad, U. P. (2017). *National Identity Politics and Postcolonial Sovereignty Games: Greenland, Denmark, and the European Union. Monographs on Greenland* (353). Viborg: Museum Tusculanum Press.

Geertz, C. (1998). Deep hanging out. *New York Review of Books*, 45(16): 69–72.

Gesink D., Rink, E., Montgomery-Andersen, R., Mulvad, G., and Koch, A. (2010). Developing a culturally competent and socially relevant sexual health survey with an Urban Arctic community. *International Journal of Circumpolar Health*, 69(1): 25–37.

Greenhough, B. (2011). Assembling an island laboratory. *Area*, 43(2): 134–138.

Hansen, K. G. (2017). *Fra passiv iagttager til aktiv deltager. INUSSUK—Arktisk for- skningsjournal*, 1. Nuuk: Greenland Self-Government/Atuagkat.

Haraway, D. (1988). Situated knowledges: The science question in feminism and the privilege of partial perspective. *Feminist Studies*, 14(3): 575–599.

Haraway, D. (1997). *Modest_Witness@Second_Millennium. Female-Man©_Meets_OncoMouse^{TM}*. London: Routledge.

Haraway, D. (2016). *Staying with the trouble: Making kin in the Chthulucene*. Durham, NC: Duke University Press.

Hayward, S. and Cassell, C. (2018). Achieving critical distance. In *The SAGE Handbook of Qualitative Business and Management Research Methods* (361–376). London: SAGE Publications.

Heine, M. K. (2013). No "museum piece:" Aboriginal games and cultural contestation in Subarctic Canada. In C. Hallinan and B. Judd (Eds), *Native games: Indigenous peoples and sports in the post-colonial world. Research in the Sociology of Sport, 7* (1–19):1–20.

Hicks, J. and White G. (2015). *Made in Nunavut: An experiment in decentralized government*. Vancouver, BC: UBC Press.

Interagency Arctic Research Policy Committee (2018). Principles for Conducting Research in the Arctic. Retrieved from www.iarpccollaborations.org/principles.html.

International Arctic Social Sciences Association. (2018). Request for Comments forwarded by e-mail from the American Interagency Arctic Research Policy Committee, August 30.

International Association for Public Participation. (2019). IAP2 Spectrum. Retrieved from https://iap2canada.ca/Resources/Documents/0702-Foundations-Spectrum-MW-rev2%20(1).pdf.

Latour, B. (2005a). Critical distance or critical proximity? A dialogue in honor of Donna Haraway (Unpublished manuscript). Retrieved from www.bruno-latour. fr/sites/default/files/P-113-HARAWAY.pdf.

Latour, B. (2005b). *Reassembling the Social. An Introduction to Actor-Network Theory*. Oxford: Oxford University Press.

Law, J. (2004a). *After Method: Mess in Social Science Research*. New York: Routledge.

Law, J. (2004b). Matter-ing: Or how might STS contribute? Centre for Science Studies, Lancaster University. Retrieved from www. heterogeneities. net/publications/Law2009TheGreer-BushTest.pdf.

Law, J. (2009). The Greer–Bush Test: On politics in STS. Retrieved from www. het erogeneities.net/publications/Law2009TheGreer-BushTest.pdf.

Law, J. and Singleton, V. (2005). Object lessons. *Organization*, 12(3): 331–355.

Loukacheva, N. (2007). *The Arctic Promise: Legal and Political Autonomy of Greenland and Nunavut*. Toronto, ON: University of Toronto Press.

Mills, C. W. (1959). *The Sociological Imagination*. London: Oxford University Press.

Pedersen, B. K. and Rygaard J. (2003a). "De Forbudte Trin" i Forskning og Undervisning. Overskridende Metoder. In *Grønlandsk Kultur- og Samfundsforskning*. Nuuk: Ilisimatusarfik/Atuagkat.

Pedersen, B. K. and Rygaard J. (2003b). Grønlandske unge mellem tradition og globalisering. In H. Helve (Ed.), *Ung i utkant: Aktuell forskning om glesbygdsungdomar i Norden* (339–357). Copenhagen: Nordic Council of Ministers.

Petersen, M. K. and Ren, C. (2015). "Much more than a song contest": Exploring Eurovision 2014 as potlatch. *Valuation Studies*, 3(2): 97–118.

Ren, C., Gad, U. P., and Bjørst, L. R. (2019). Branding on the Nordic margins: Greenland brand configurations. In *The Nordic Wave in Place Branding* (160–174). Cheltenham: Edward Elgar Publishing.

Ren, C., Mahadevan, R. and Madsen A. K. (2016). Valuation and Outcomes in the Arctic Winter Games 2016: Contributions from Research. Department of Culture and Global Studies, Aalborg University. Retrieved from: https://vbn.aau.dk/ws/p ortalfiles/portal/243377064/Report_AWG_AAU.pdf.

Ren, C. and Mahadevan, R. (2018). "Bring the numbers and stories together": Valuing events. *Annals of Tourism Research*, 72: 75–84.

Ren, C., Petersen, M. K., and Dredge, D. (2015). Guest editorial: Valuing tourism. *Valuation Studies*, 3(2): 85–96.

Rygaard, J. (2010). "Proxemic Nuuk:" Town and urban life with Nuuk as example. In K. Langgaard et al. (Eds), *Cultural and Social Research in Greenland—Selected essays 1992–2010* (249–302). Nuuk: Ilisimatusarfik/Atuagkat.

Sejersen, F. (2015). *Rethinking Greenland and the Arctic in the Era of Climate Change: New Northern Horizons*. London: Routledge.

Sermitsiaq. (2012). *Kan vi bruge pengene bedre?*April 6.

Shadian, J. M. (2014). *The Politics of Arctic Sovereignty: Oil, Ice and Inuit Governance*. London: Routledge.

Skille, E. Å. (2013). Lassoing and reindeer racing versus "universal" sports: Various routes to Sámi identity through sports. In C. Hallinan and B. Judd (Eds), *Native Games: Indigenous Peoples and Sports in the Post-colonial World. Research in the Sociology of Sport, Vol. 7*:1–19, 21–41.

Thisted, K. (2011). Greenlandic oral traditions: Collection, reframing and reinvention. In K. Langgaard and K. Thisted (Eds), *From Oral Tradition to Rap: Literatures of the Polar North* (63–117). Nuuk: Ilisimatusarfik/Atuagkat.

Thisted, K. (2017). The Greenlandic Reconciliation Commission: Ethnonationalism, Arctic resources, and post-colonial identity. In L. Körber, S. MacKenzie, and A. S. Westerstahl (Eds), *Arctic Environmental Modernities from the Age of Polar Exploration to the Era of the Anthropocene* (231–246). Basingstoke: Palgrave Macmillan.

Thomsen, R. C., Ren, C., and Mahadevan, R. (2018). We are the Arctic: Identities at the Arctic Winter Games 2016. *Arctic Anthropology*, 55(1): 105–118.

Thomsen, R. C. and Vester, S. P. (2016). Towards a semiotics-based typology of authenticities in heritage tourism: Authenticities at Nottingham Castle, UK and Nuuk Colonial Harbour, Greenland. *Scandinavian Journal of Hospitality and Tourism*, 16(3): 254–273.

Vikkelsø, S. (2007). Description as intervention: Engagement and resistance in actor–network analyses. *Science as Culture*, 16(3): 297–309.

Wiita, A. (2006). The culture of community-based research and a borderless north. *Fourth Northern Research Forum Open Meeting*. Retrieved from www.rha.is/static/files/NRF/OpenAssemblies/Oulu2006/project-community_wiita_second.pdf.

8 Life Mapping

A collaborative approach to tourism collaboration in Greenland

Daniela Chimirri

Introduction

Within the field of tourism, collaboration is regarded as key to producing and offering products and services. The fragmented nature of tourism and its "assembly process" characteristics make it imperative for tourism actors to work together despite their diverse interests and goals (Ladkin and Bertramini, 2002).

> My focus is on the area and the local actors. I just don't have the resources and time to contact international actors. It isn't that we rule out working with internationals. We also need them … However, there is a limit on how much I can do with the resources and time I have. Ultimately, we need others to help create experiences for the tourists.
> Research participant, data from own PhD research project, April 2018

This example from Greenland demonstrates how actors have different access and control over resources. There are very few situations wherein a single tourism actor exclusively holds the control over all of the individual components and/or the decision-making processes in order to build the political framework for producing/creating as well as finally delivering the tourism products and services.

However, the acknowledgment and recognition of the significance of working together obscures the fact that articulating collaboration as an essential element in the daily operation of tourism businesses is not easy. Drawing on research among Greenlandic tourism actors, this chapter explores how the "life mapping" can contribute to our understanding of collaborative practices in tourism. The life mapping exercise takes place through a diagramming process during "classic" qualitative interviews. I am using the collaborative life mapping method in order to collaboratively explore the research participants' collaborations in the Greenlandic tourism landscape. My very research subject is the concept of collaboration. In terms of understanding how tourism entrepreneurs in Greenland work together and how this might affect the development of tourism, I must be part of this process of mutually exploring and creating an understanding of the collaborative land-scape of this Arctic destination. Collaboration is not my sole research subject, and tourism entrepreneurs are not only research objects representing carriers of knowledge. We are collaborators (Ren, Jóhannesson, and Van der Duim, 2017). I therefore understand my research process as a bottom-up and co-creating approach aimed at fostering an understanding of the collaborative practices from within.

This chapter argues that the combination of verbal description and the visual illustration of collaborations allows the research participant and researcher to co-create an understanding and new knowledge of practices. It initiates reflections on mundane daily practices through dialogue and visuali-zation, potentially activating new thoughts and considerations about the the-oretical concept of collaboration, as well as collaboration-inspired research approaches. The chapter ends with reflections on the outcomes of this co-construction method from a conceptual, methodological, analytical, and epistemological perspective, advocating for more collaborative approaches to research in the Arctic.

A call for more collaboration in the Arctic and in Greenland

Agreeing with Pain and Francis (2003), there is a need to turn toward more collaborative research that generates "bottom–up knowledge which is pro-duced in a rigorous ethically acceptable way—in other words, to have real impacts for those we study beyond academic articles and conference papers" (2003: 47). Holm et al. (2011) underline this need to create a research practice in the spirit of working collectively and, therefore, in the context of this book, to connect local communities in the Arctic with ongoing research.

As also normalized in other parts of the Arctic, "hit and run" behavior among scientists appears to have become common practice in Greenland. Such research practices continue despite becoming much less accepted in recent decades, and research often takes place past local communities. The following example from Greenland illustrates this perceived failure of research in the Arctic (interview excerpt, research project in 2016–17; see Ren and Chimirri, 2017):

RESEARCH PARTICIPANT: [1] Then [referring to the building of a visitor center in the North of Greenland functioning as meeting place for locals, tourist information office and research facility], we will have a closer working relationship to them [the researchers visiting the area for study purposes]. Because at the moment we don't usually see them unless we meet them. Like, they stay in the hotels … then they just travel back to [wherever they came from].

RESEARCHER: So, the physical building also offers you a way to collaborate closer with others and to be able to exchange knowledge?

RESEARCH PARTICIPANT: Yeah, for sure.

RESEARCHER: When scientists are visiting, don't they contact you? I assume in your position—you probably know many people who have experience with the ice cap and would love to share their knowledge with others. That might also help scientists, no?

RESEARCH PARTICIPANT: Yes, for sure. I heard a while ago that there's a couple from XX.[2] Last summer, they had their 10th anniversary of research here in Greenland. The whole June, they study the ice, the sea currents, and the temperature. And we didn't know about them.

The respondent articulates the sense of being uninformed about and excluded from research taking place in and concerning them as Arctic inhabitants. Throughout the interview, the research participant also emphasized their interest in working collectively. Working collaboratively would give local communities a stronger voice in the public and academic discourse of Arctic-related issues and challenges. He believes that such approaches benefit and give back to the local communities while at the same time enriching the quality and relevance of research.

Following this statement, the Arctic communities appear to be much talked about—but less talked *with* in many cases. This perspective is expressed even more explicitly in the following excerpt of an interview conducted during my own research project in March 2019. It points to the growing concerns of Arctic communities of not being knowledgeable about and included in ongoing research.

RESEARCHER: The Arctic regions attract more and more interest from different sides. A lot of challenges and issues are being discussed in the media.

RESEARCH PARTICIPANT: In a way, I think that's harmful. It's also disrespectful when we look at how Arctic people have been living with change all the time. We're not dying out, we're not disappearing. We're just adapting.

RESEARCHER: In your opinion, is it a problem how the Arctic is represented, for example in the media?

RESEARCH PARTICIPANT: In a way, yes. I think it's important to talk about climate change, for example, but I think it's equally important to acknowledge that people living here were always able to adapt or they would have died out. If we look at the history of Arctic expeditions, they always forget to mention the local helpers. They are usually a footnote, if mentioned at all. In many aspects, the West has been really good at forgetting the indigenous population of the Arctic.

RESEARCHER: Do you mean that they don't have a voice? That they should be more in the picture, so to speak?

RESEARCH PARTICIPANT: Yes, for sure. A very good publication came out a few years ago: "The meaning of ice." It's about hunters from Alaska, Canada, and North Greenland talking about climate change and their relationship to ice. That's one of the human perspectives: the perspective of the Arctic people … it's also important to consider the people living in and off the ice. It's one of these narratives that the Western world has been good at forgetting.

Based on these arguments and theoretically inspired by Bent Flyvbjerg's work with Aristotle's concept of *phronesis* [3] (Flyvbjerg, 2001, 2004, 2005, 2006), as Arctic researchers, we must follow a research approach that generates "experience in context as the most appropriate means of generating knowledge that matches social priorities and can contribute to public debate" (Thomas, 2012: 12). By engaging and involving research participants in, continuously communicating research results to, discussing with, and incorporating feedback from the public, the generated understanding and knowledge build the ground for the research "at place" (Flyvbjerg, 2006). As a kind of bottom-up, collaborative approach to research, doing it this way serves as the eyes and ears of the researchers' ongoing efforts to understand the present and to deliberate about the future from the perspective of the "object under study" (Flyvbjerg, 2006; Pain and Francis, 2003).

Shaping research from below

Following a phronetic research approach by focusing on practical activities and the knowledge of everyday life situations, this approach in a previous research project showed that tourism actors in the Greenland tourism landscape regard collaboration as significant and acknowledge the necessity of engaging in actions and mutually working together with others (Ren and Chimirri, 2017). "It's through the cooperation, the networks, and one's relationship to others that things happen. Working together, we can find ways to do things differently, better, to make things happening" (Research participant, research project in 2016–17; see Ren and Chimirri, 2017).

Tourism actors in Greenland have different ways of dealing with existing conditions, possibilities, and challenges in their daily practices. Their realities are highly complex, diverse in nature, very dynamic, and highly unpredictable, owing to multiple and hardly controllable circumstances, including distance, a harsh climate, and a dispersed population (Ren and Chimirri, 2017, 2018). What the tourism actors have in common is their constant awareness of how they are bound to work together. In the words of one research participant: "[It] is a bit difficult … we have a limited budget. The things we would like to do cost money and we also need the expertise. Collaboration helps and we're always looking out for it. We're open to any kind of collaboration, really" (Research participant, data from own PhD research project, April 2018).

While this seems to be the case, the awareness and testimonies about the "greatness" of collaboration cannot hide the fact that this considerable and essential element of everyday life practices in the Greenlandic tourism landscape is not easy for tourism actors to articulate. It is difficult to grasp for research participant and researcher alike. On the one hand, statements like "when we work together, it's really good for us" (Research participant, data from own PhD research project, April 2018) illustrate the value that tourism actors attribute to collaboration. On the other hand, by saying "we don't really think about it—we just do it" (Research participant, research project in 2016–17; see Ren and Chimirri, 2017), actors struggle to articulate the inherently complex nature of collaborative actions. It feels difficult to actively put words on "collaboration." The inherent meaning nested in these interview statements illustrates the challenge of articulating what seems to be so normal in the actor's daily routines.

Methodological tool to explore collaboration

In theoretical terms, there is also a lack of analytical tools for grasping the term "collaboration," even though it is used extensively nowadays. Consensus seems to exist regarding the increasingly important role of collaboration between and across public and private, nonprofit and profit sectors, owing to that which Gray (1989) calls "the need to manage differences" (p. 1). However, the concept incorporates and carries very different and diverse meanings, depending on the context in which it is used (Morris and Miller-Stevens, 2016b). I therefore argue that it is necessary to find and use methodological tools that enable us—the research participant and researcher—to explore this mundane activity in order to move toward a less superficial, theoretical articulation of the term and a more concrete and practical understanding of collaboration.

The following section of this chapter presents the "life mapping" method (inspired by Marschall's work, 2013, 2017) as a means to collaboratively explore these daily practices and investigate how collaboration takes place in the Greenland tourism landscape. After introducing this method, I demonstrate how I have applied it in my own research. It illustrates how researcher and research participants can collaborate on exploring the collaboration concept by co-creating life maps in interviews.

Life mapping: Unfolding everyday life practices through the co-construction process of participatory diagramming

In tourism studies, Barry (2017) examines the challenges to articulating and documenting everyday practices from the perspective of the tourist:

> Everyday tourist practices … are often so subtle, momentary and ordinary, yet form a significant part of a tourist's daily routine. They demand a considerable amount of time, attention and practiced negotiation, but can be difficult for tourists to articulate and reflect on and for researchers to document. (p. 328)

There is a need for creative and collaborative research approaches to engage in and find new and additional ways of capturing and documenting everyday life practices in tourism research. Such new approaches include possible additional "navigational tools than the spoken language" (Marschall, 2013: 8). In this line of argumentation, "life mapping" is regarded as a dialogical method of co-constructing data as part of a participatory diagramming approach (Kesby, 2000; Kesby et al., 2005; Literat, 2013; Pain and Francis, 2003) and represents an alternative and yet collaborative approach to data generation within the field of Arctic research.

Participatory diagramming refers to a set of methods and visual techniques ranging from making sketches, drawing cartoons and transects, mapping, and compiling charts, diagrams, and matrixes (Kesby et al., 2005). Based on my own PhD research project, I introduce the method of life mapping in this chapter. The description of the mapping process and the illustrated examples come from my fieldwork, which I conducted in South, West, and East Greenland between June 2017 and July 2019. The research participants and I co-created 42 life maps during a total of 40 interviews in the towns and settlements of Qaqortoq, Nanortalik, Narsarsuaq, Nuuk, Maniitsoq, Kangerlussuaq, Sisimiut, Kulusuk, and Tasiilaq.

When looking through the lenses of cartography, mapping refers to the visual representation of "looking from a satellite and getting the overall picture of continents, countries, and oceans" (Marschall, 2013: 8). In connection with the research focus of my research project, however, the culmination and expression of the diagramming process are "life maps" capturing everyday life practices in the form of visual representations (Marschall, 2013) of collaborative activities and formations of collaboration in the Greenlandic tourism landscape.

The following excerpt from an interview situation (data from own PhD research project, April 2018) picks up on these aspects. It illustrates the struggle of research participants to articulate their daily life practices, which affects and hinders the process of unfolding the concept of collaboration. It also shows how the mapping process helps to create an atmosphere for dialogue between researcher and research participant leading to the joint investigation and unfolding of information and insights, valuable for both parties in terms of knowledge creation.

RESEARCHER: I'd like to try something if you're up for it. I'm trying to get a picture of how collaboration works here in Greenland, starting here in South Greenland. How do you work together? Or why do you not? How would you visualize the collaborations with others? Could you draw or write the collaborations you have on this? [hands a pen and piece of paper to the interviewee] Can we try that?

RESEARCH PARTICIPANT: Hmm, well, for us it's … we work with everybody somehow. I don't know, it isn't … I'm actually not sure how to … what do you want me to draw?

RESEARCHER: Whatever comes to mind when you think about collaboration. Everything and everyone you can think of. There's no right or wrong.

RESEARCH PARTICIPANT: OK. We work a lot with the hotel, because all of our clients are here at least one or two nights. On the other side of the fjord, there's a farm/guesthouse. It's just on the other side. Then, you can walk to the other side of the fjord, where there's also another farm. And then you have another one we work with. It's good now. It has been difficult though, but now I think something is developing.

RESEARCHER: So these ones [pointing at the map], they're offering accommodation and the others horse riding, or?

RESEARCH PARTICIPANT: Yes. We have the tourists, and they make the offers. Ahh, there's also another farmer we work with a lot, just opposite the ice fjord.

RESEARCHER: How does the collaboration work? Do you have personal contact or is it more via email?

RESEARCH PARTICIPANT: A lot via email, but we also meet sometimes. I almost forgot, we have the restaurant in Igaliku. It's a hard place to run, because it's only open in the summer and every year it's very difficult to find someone. It's only open 3–4 months and it's hard work. It's very difficult to find someone.

RESEARCHER: Would you say that finding employees is one of the biggest challenges for tourism in Greenland?

RESEARCH PARTICIPANT: Yes, I think so. Definitely for us ... For us, it's getting too hard, because of this. And that's why I also—we have all these people we're working with [pointing at the map]. We have some that we really work a lot with, because they get involved and we can rely on them.

(Interview excerpt, own research data, April 2018)

When talking about collaboration as part of daily life practices, the emerging maps are not "a static representation of information, but rather a set of relations that emerge through events and processes" (Barry, 2017: 331). Through the joint process of life mapping, maps are therefore continuously talked about and discussed and work "as 'papers in progress' between researcher and research-participant" (Hviid and Beckstead, 2008: 161). Hence, the life maps can (but do not necessarily have to) be visual representations of the overall picture of the collaborative landscape in Greenland at the time of their creation during the diagramming process. They can depict the existing situation and represent a snapshot. They can also comprise desired future formations and collaborations. In other words, the life maps might also be a mirror for possible collaborative activities and a potential collaborative tourism landscape, which is not yet in place.

What is important is that the researcher and research participant co-construct the process. Such an approach aims to create an environment in which actors feel comfortable to share and exchange knowledge and experiences, but also to reflect, enhance, and reconsider views and practices (Hasse and Milne, 2005). As a non-verbal and visual medium of representation, the mapping exercise provides researcher and research participant with an opportunity to "talk." They can express themselves in a more "pleasant" way of communicating and/or might be able to put "words" onto aspects, such as everyday life practices, that are otherwise difficult to articulate.

In the context of Greenland, the interviewees were mainly Greenlandic and some of them Danish-speaking (referring to the language as their mother tongue). Neither Greenlandic nor Danish is my mother tongue. This precondition can be problematic when aiming to reach and talk to members of the local communities. First, I cannot expect research participants to speak my own mother tongue, which is German. Second, English is the third language for most of the interviewees with whom I was in contact, after Danish (as the second language taught from first grade, Statistics Greenland, 2018). And thirdly, it is also foolhardy to assume that it is possible for me to learn Greenlandic. I therefore required additional "navigational tools other the spoken language" (Marschall, 2013: 8), and mapping became a way to mitigate the dependence on linguistic proficiency—for both sides.

As noted by Brennan-Horley et al. (2010) in an article on mapping technologies as forms of ethnographic methodologies, such processes activate "a different attitude to the interview situation on the part of the participants than a more straightforward and expected question-and-answer format" (p. 96). In my own experience, the "creative experiment" of life mapping changed the atmosphere of the interview in many cases from a more formalized setting in the beginning to a more relaxed tune once the life mapping had started (see below, own research data, April 2018).

RESEARCHER: Anyway, I'd like to try an experiment and I hope you're up for a bit of a creative task?
RESEARCH PARTICIPANT: Like—you're creative or I'm the creative one?
RESEARCHER: You're the creative one.
RESEARCH PARTICIPANT: Oh, uh, OK, well … I'm short on coffee [laughing], but I'll give it a shot.
RESEARCHER: Great. If you think about the collaborations you have, how would you visualize them?
RESEARCH PARTICIPANT: How to visualize…? Hmmmm, like drawing?
RESEARCHER: Drawing, writing, doing a sketch, anything. Whatever comes to mind.
RESEARCH PARTICIPANT: OK, here we go [starts to draw].

The participatory diagramming holds "the inherent potential of painting a more nuanced depiction of lived realities, while simultaneously empowering the research participants and placing the agency literally [also] in their … hands" (Literat, 2013: 12). The manner in which this process was applied is introduced in the following section.

The process of co-creating 'life maps'

The participatory diagramming process took place during "classic" qualitative interviews, and an interview guide was developed to outline guiding questions. These guiding questions mainly function as "interview openers." Prior to the interview, the research participants did not know that there would be a diagramming exercise, and it was unknown how the respective research participants would perceive the creative task of drawing. As a "cold start" might have overwhelmed the research participants, it seemed reasonable to start the interview with a few opening questions. These questions were of a general nature regarding the participant's workplace, tasks, and so forth.

I decided on the moment to start with the diagramming exercise depending on the situation. In some instances, the research participant already started talking about the aspect of collaboration, and the researcher seized on this to start the exercise (as illustrated in the citation below, own research data, April 2018).

RESEARCH PARTICIPANT: Exactly. More and more Greenlandic tour operators are coming. XX, for example, from Qaqortoq and also Narsaq. We work together with XX and YY.

RESEARCHER: As we're already getting into the collaborations you have with others, I'd like to do a small experiment. How would you visualize the collaborations you have? Could you draw them on this piece of paper?

RESEARCH PARTICIPANT: Does it need to be a nice drawing?

RESEARCHER: Not at all. You can draw or write. Whatever you like. I'm trying to unfold how actors work on the ground. Who you work together with and that kind of thing. You already mentioned quite a lot of partners you work with and others you'd like to work with in the future.

In other cases, I chose a moment that fit the flow of conversation. In the following excerpt, for example, the research participant and I were talking in general terms about the tourism landscape in the area, and the research participant's statement initiated the starting of the diagramming:

RESEARCHER: It seems to me that there is a vivid discussion on how to develop tourism in the future.

RESEARCH PARTICIPANT: Yes, for sure. We don't know the future yet, but we talk together about it and work together. So things might happen.

RESEARCHER: That leads me to my next questions on who you work with and how. And how would you visualize it on this piece of paper?

RESEARCH PARTICIPANT: We work with everybody. In a situation like now, where tourism is growing, it's very positive. Everybody is getting more. In a situation where it's the other way around it would be different. But right now—yesterday we had XX coming by and presenting what they do. We work with them. Then we had YY dropping by. We're working close with their competitor, but we also use their services. We work with everybody. That's the tourism landscape in Nuuk [pointing at the finished drawing].

RESEARCHER: Where are you?

RESEARCH PARTICIPANT: We're one of the bubbles. We're here and work together with everybody.

RESEARCHER: So everyone is in this one bubble and works together—How?

(Interview excerpt, own research data, March 2019)

Aiming to keep the process as open as possible, the research participants were asked: "How would you visualize the collaborations you have?" This initial question, aiming to start the diagramming process, led to diverse and different reactions and, as shown in section four, to the emergence of very different and diverse life maps. However, most of the research participants were in fact surprised about the task and started in similar ways—as this one excerpt from one interview situation aptly demonstrates (own research data, April 2018):

RESEARCH PARTICIPANT: OK. Well ... interesting. I'll try.

RESEARCHER: You're free to do whatever you want. It could be organizations, institutions, businesses, municipality, the choir, etc. Anyone you work with, partners you would like to have, and so on.

RESEARCH PARTICIPANT: OK. Should I draw or write?

RESEARCHER: Whatever you prefer. That's up to you.

During the drawing process, the researcher asked clarifying questions (see the two interview examples below). Decoding the drawings was sometimes difficult, and the questions aimed at getting information about what was drawn (as the drawings were sometimes not named/labelled and/or such labels were sometimes illegible).

Example 1:

RESEARCHER: Where are you [pointing at the drawing]?

RESEARCH PARTICIPANT: We're one of the bubbles. We're here and work together with everybody.

RESEARCHER: So everyone is in this one bubble and works together—How?

Example 2:

RESEARCH PARTICIPANT: OK. Well, when we write the museum here. Then we have the tourism industry here.

RESEARCHER: Who is the tourism industry for you?

RESEARCH PARTICIPANT: Right now, it's the local tourism actors. All the ones working with tourism.Some research participants found it more challenging than others to draw life maps of their work, as indicated in the following:

RESEARCHER: I'd like to try an experiment with you. I'm working on mapping the collaborative landscape in Greenland. And I'd like to ask you to draw your collaborations. How would you visualize your collaboration?

RESEARCH PARTICIPANT: Well, that's actually going to be very simple. There's just the hotel and then some others. Here in the local area, there's just XX. They're the only ones we directly work with.

RESEARCHER: So when we have the hotel on this piece of paper, what then? I'm trying to motivate you to draw right?

RESEARCH PARTICIPANT: Yes, yes. I know [laughing and starting to draw]. The drawing process and the researcher's questions to the emerging life maps prompted the research participants to (re-)think their drawings and verbalize thoughts regarding their own (in some instances missing) position on the maps, connections, and relationships that they have or would like to have with other actors.

After the presentation of the process of creating life maps, the following section will address the everyday practices that these maps actually reveal.

Life maps—What do they show?

As part of my own research project, 42 life maps were co-created with multiple tourism actors in fieldwork in different parts of Greenland in the period June 2017 to July 2019. An exhaustive presentation of all of these created maps would exceed the scope of this chapter. Three maps were therefore chosen and will be presented in the following. These maps are briefly introduced and described to present "what they show" in order to explore the collaboration concept.

Every life map is a symbolic representation presenting us with information about how the research participants view collaboration, and each of them provides very particular insights into collaborative practices. Each individual research participant chose a different way of "how to visualize" their collaborations. The visualization of these mundane practices and collaborative instances are crucial in connection with the verbal interview data.

The following interview excerpt (own research data, April 2018) demonstrates three essential aspects when talking about the concept of collaboration: the perceived lack of collaboration (despite it actually existing!), the importance of infrastructure in the context of collaboration, and organizational structures as agents for collaboration.

RESEARCHER: I'm trying to find out what collaborations you have—who do you work with? How does it work? And so on.

RESEARCH PARTICIPANT: Well that will be very short. There is **no cooperation**. I have been in the tourism business for 20 years now. I have my own business here. In the beginning, there were a lot of possibilities, a good infrastructure. There were passenger boats to all the settlements. Flights three times a week … We had a **good infrastructure**, which also meant that we had **good possibilities to cooperate** … we all united as tourism destination … We formed [the DMO[4]]. We built up a **good**

organization ... [I]t was a well-functioning organization ... We **worked together regarding developing products and providing services** to cruise ships. We represented the destination abroad to different companies to sell our products ... It was a good organization and close cooperation. In 2009, [it] was restructured ... we didn't receive support and we couldn't exist anymore. There were different reasons, but one of the major reasons was the missing funding. We weren't able to proceed. We were left alone. ... I'm the only one left that works with tourism fulltime, year-round.[...]

RESEARCHER: As we're already talking so much about it, could you draw your collaboration for me?

RESEARCH PARTICIPANT: I write XX. I have no collaboration with YY. We were working together before, but ... [we] don't collaborate anymore. In relation to XX, I also have an incoming agency ... I sell them tours and accommodations. And in order to do that, I **work together** with XX a lot. We help each other if necessary ... With ZZ [pointing at a circle outside with no link to the research participant in the middle] there's no cooperation ... It doesn't work yet, but it's necessary if we're going to make a good tourism destination ... They have to learn to cooperate. They're thinking too much about making money, to keep [business] for themselves and not sharing with others. The spirit of cooperation—they don't have it.

The three aspects articulated by this research participant are also reflected in other co-created life maps. The chosen three maps illustrate these aspects separately in the following sections, whereas the first life map (Figure 8.1) is co-created during the interview of this specific excerpt.

The life map in Figure 8.1 reveals a collaboration between some tourism actors and that the research participants do not work together with a few actors. These circumstances are marked by two-sided arrows in the map when considering existing collaborations. The actors with whom the research participant does not work together are drawn on the side without any links to other actors in the map.

The fact that there is some collaboration among a few actors and that there are also actors who do not work together with others hardly seems surprising, as this would seem normal when thinking about collaboration in general terms. However, this specific life map is the map from the interview excerpt above. It explicitly reveals how *collaborative structures exist between the tourism actors, even though the research participant explicitly stated in the beginning of the interview that there is no collaboration at all.*

This example shows the significance of using life mapping to explore collaboration. As the research participant denied the existence of collaboration early in the interview, the question becomes whether a classic interview would have turned up elements of collaboration. It also shows how difficult it is for tourism actors to articulate their daily practices in detail and that the diagramming process and the emerging life maps help to initiate reflections.

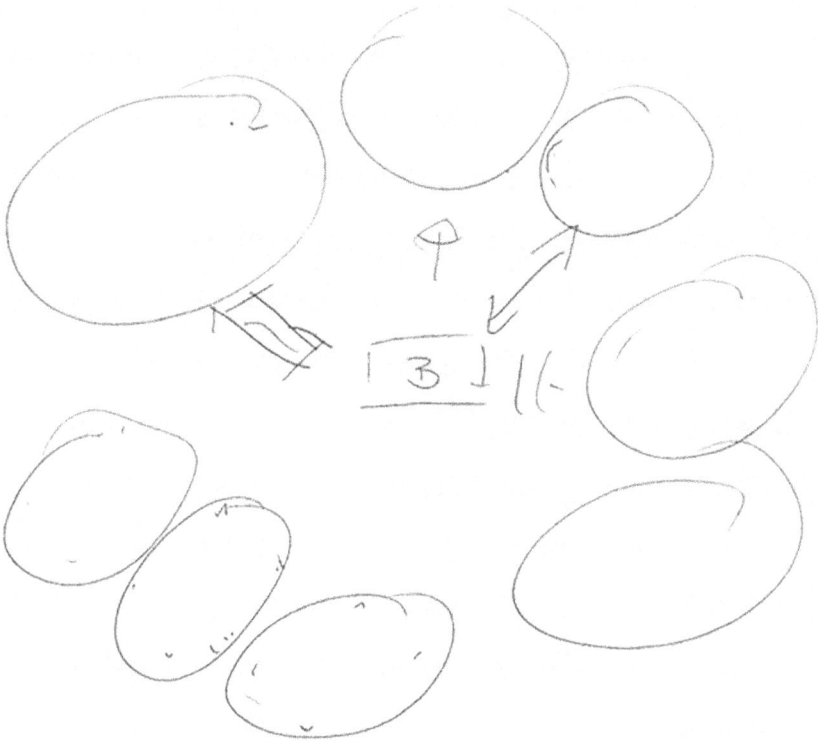

Figure 8.1 Life map visualizing (non-)collaborations (April 2018)

Returning to the interview excerpt, the research participant mainly explains the lack of collaboration with the deficiency of physical infrastructure (in this case, in terms of transportation) as well as in the loss of the local DMO as an organizational structure. In the Arctic, the collaboration concept is heavily shaped by materials as well as structural aspects. The circumstances in Arctic destinations regarding weather conditions and infrastructure influence how actors conceive of collaboration. In the ongoing public and political discourses in Greenland, people "just" talk about these circumstances, but the following life maps (Figures 8.2 and 8.3) show the importance of infrastructure and the role of organizational structure as agents within the collaboration context.

The life map in Figure 8.2 illustrates the significance of infrastructure. Here, the research participant depicts both human actors and non-human actants as part of the physical and digital infrastructure creating the collaborative landscape at place.

RESEARCH PARTICIPANT: OK. Here's my smartphone, which is important for me to communicate with others. To get a message out and to reach as

Figure 8.2 Life map of actors and actants as part of the physical and digital infra-structure (April 2018)

many people as possible in Greenland, I put it on Facebook. Then we have my colleagues here [pointing to the single person on the paper]. They're a huge source of knowledge. Then obviously I have contact with all kinds of people that do things … I think what's important when we talk about working together … I think our main issue in this is infra-structure—or the lack of it. When you're in a small town, you're like on an island. We need to be connected.

This life map and the supportive interview data render it obvious how material also matters. The human interaction is in the center of the map (with the table with two people and the single person), and the research participant draws a social situation with speaking bubbles and the word "coffee." This clipping of the map illustrates sociality. By placing it in the center of the map, the research participant emphazises how the inter-personal social relationships are very important for her when talking about collaboration. By including physical objects in the map, however, she also makes clear that material artifacts are necessary for it to be even possible to work with others; and that this aspect is also credited high significance.

The physical infrastructure of actants—including telephones, checklists, and the archives as way of documentation and the digital infrastructure of internet connection—are equally important to the human interrelations in collaborative interactions. This life map therefore illustrates the crucial relation of non-human actors with human actors as part of enabling collaborative activities.

The combination of the physical and digital infrastructure also allows for organizational structures to form and to operate, such as DMOs as an element in the tourism landscape. The research participant in this interview excerpt emphasizes the importance of "a well-functioning organization [for working] together" (Research participant, own research data, April 2018). In Greenland, the tourism structure and how it is organized varies from region to region, and questions on where to go and how to do tourism are constantly debated amongst Naalakkersuisut[5], representatives of diverse organizations (public and private) and local tourism entrepreneurs in different settings from public hearings, conferences, seminars, and workshops.

Undeniably, "there are different ways to do it. What's important is [the question on] who takes care of the development ... Who coordinates and launches [ideas and proposals]? How do we want to work together on this?" (Research participant, data from own PhD research project, April 2018). The last life map in Figure 8.3 illustrates how the DMO as organizational structure is seen as helper/mediator in the organization and facilitation of tourism and its diverse actors.

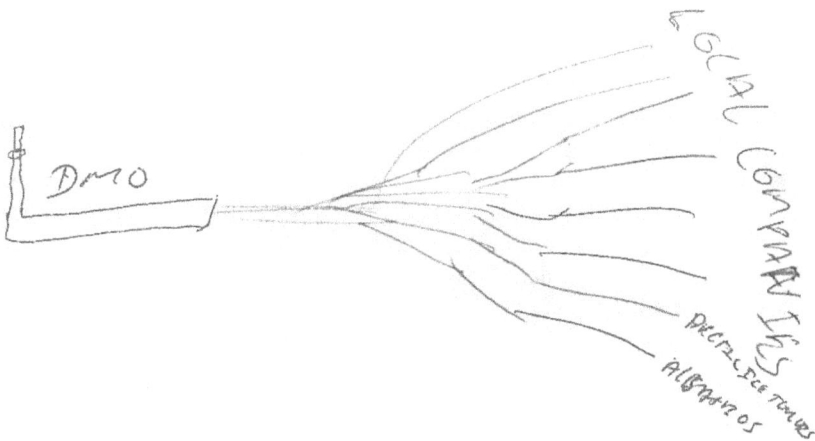

Figure 8.3 Example of a life map of collaborations as metaphor—the dog sled

RESEARCHER: That actually leads me to my next question, and I'd like to try a little experiment. Who do you collaborate with, and how would you visualize these collaborations?

RESEARCH PARTICIPANT: We just talked a bit about this actually in Ilulissat, and we had a difficult time trying to figure it out. You can think about it along the lines of a dog sled ... [There are] many small local businesses. And they need a lot of knowledge. That's mainly what we do here. Us in the council ... That's why I think it is us on the sled—because we can help them go in the direction they want. The example doesn't really work that way, because all of them don't necessarily move exactly the same way. That's something we need to help them with and make them aware of. If you want more tourism, you need to understand that we need to move together toward this goal. And even though most of the companies are competitors, they also need to understand that they need to work together to get more tourists. A lot of them don't offer accommodations, and those who do offer accommodations don't necessarily offer tours. And some of them who offer tours, like dog sledding, need others that also do dog sledding—because if a group of 30 comes along, then they need a lot of dog sleds ... They always need to work together. And when they do, it actually works better. It isn't me deciding where we're going. I need these guys. If they don't want to be part of it, then they start going in all kinds of directions—without really moving the sled.

Here, the DMO as organization is considered as agent, enabling collaboration by aiming to unite the diverse tourism actors in the destination. As illustrated in the life map and supported by the interview data, the DMO takes an active role. The DMO functions as an agent, mediating between actors and bringing them together to collaborate. This organization also takes the position of the musher by sitting on the dog sled (see Figure 8.3) and steering the tourism actors. It actively shapes the tourism landscape. Another research participant working for a DMO explains how:

> [i]t's important to note that [tourism actors] need knowledge and resources, which are very much based on the language skills needed to interact with tourists and the creation of the produce. It's [also] important to know what the [tourism actors] need to give the tourist a good experience and the other way around. We [the DMO] get to play a part in that. We have service courses coming up this year. We provide some language courses. [With these activities] we help actors to develop their business and create a better product.
>
> (Research participant, data from own PhD research project, March 2019)

Reflections—Outcomes for the collaboration concept

As argued throughout this chapter, exploring the concept of collaboration can be difficult due to its entanglements with practices so inherent to our daily life. The presentation of the co-created life maps of Greenlandic tourism actors reveals how the collaboration concept is highly diverse and enables us to gain a deeper understanding of collaboration.

As termed by Gray (1985, 1989; Wood and Gray, 1991) and as argued by the vast majority of tourism scholars, the collaboration concept represents a framework that helps to create an understanding of how and why actors meet and act jointly (Morris and Miller-Stevens, 2016a). The reason for doing so is seen in the process of trying to "constructively explore ... differences and search for solutions that go beyond [the] own limited vision of what is possible" (Gray, 1989: 5).

Based on the empirical data from my own research, I argue that this widely used theoretical framework cannot exploit its full analytical potential and create valuable knowledge for the research participants and researcher. As illustrated above, research participants understand collaboration differently. Even though they were all asked the same question and given the exact same task, different layers of information and multiple levels of abstraction become apparent through the diagramming process and creation of life maps.

Based on the empirical data and fieldwork experience, the combination of verbal and visual elements initiates reflections on practices in the researcher–research participant dialogue and the co-created visualization of life maps. This process potentially activates new thoughts and considerations and provides research participant and researcher alike with new knowledge. Depicting all of the various forms of collaborative activities and landscapes challenges the collaboration concept, as the emerging life maps literally illustrate the diversity of the concept. This casts light on its ontological complexity, which has previously received limited attention.

So what? Implications and perspectives

In terms of investigating the concept of collaboration as part of daily practices, the research process can be conducted with an analytical distance to the research subject by investigating it from the "outside." However, in order to do research meaningfully and to create knowledge derived from research matter in practice (Flyvbjerg, 2001, 2004, 2005, 2006), we researchers must rethink our ways of research by considering our own role as researchers and reconsider how we engage with local communities, our fellow scholars, and other interested groups.

One way of doing so has been presented in this chapter based on collaborative life mapping. I engaged in and aimed to create a research process that is collaborative in nature with respect to worldview, methodology, and research subject (see Figure 8.4).

Collaborative research incorporating a collaborative worldview and methodology making use of collaborative methods creates the grounds for a "fresh start" and aims to initiate the turn toward rethinking existing research practices in the Arctic by engaging Arctic communities in research

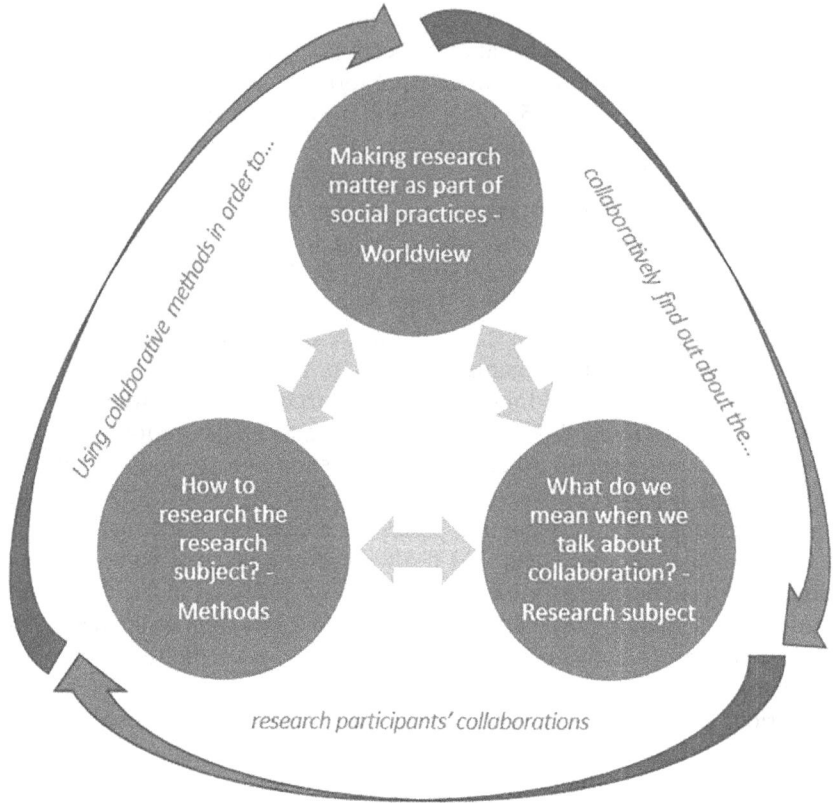

Figure 8.4 Approach to co-exploring the collaboration concept

practices and co-creating research outcomes relevant for both the local and research communities. This creates space for new ways of knowing. In my own research, life mapping stimulated new thoughts and reflections among the research participants about their own practices. This produced new knowledge on both sides of the interview table (see citations below, own research data, November 2018).

EXAMPLE 1: Now as we talk about it. By looking at it, I think...
EXAMPLE 2: I think I'm going to do this for real, this graph. You don't really think about these relations normally.

The creation of knowledge and raising of awareness can empower local actors to take an active stand in the ongoing discussions concerning tourism development and the questions "Where are we? Where are we going? And what's needed to get there?" (Naalakkersuisut, 2015). These questions are widely debated and contested by local tourism actors, who indicate that they do not feel included in the process of developing tourism in their own country, even though they represent the main actors responsible for the operational work (Ren and Chimirri, 2017). A collaborative approach potentially leads to capacity building, which is considered crucial in relation to the future tourism development of this Arctic destination and must be explored in further research.

Notes

1 The term "research participant," as used by Kesby (2000), is used throughout this chapter as it identifies the interviewee as being the knowledge carrier. The participant's experience and knowledge are valorized and the agency lies within the participant. The researcher detaches their own self from the "expert status." This relationship is characterized as reciprocal throughout the research process.
2 Anonymized, as it is irrelevant for the meaning of the statement from which country this research team came from. Similar situations apply to researchers from other countries.
3 As termed by Aristotle, "phronesis" is practical wisdom, practical judgement, common sense, or prudence. While based on Aristotle's original concept, it is more of a lived practice for Flyvbjerg (2001), which derives from a familiarity with the unpredictability and uncertainties of social practices (Schram, 2012).
4 Abbreviation for Destination Management Organization
5 The government of Greenland.

References

Barry, K. (2017). Diagramming: A creative methodology for tourist studies. *Tourist Studies*, 17(3): 328–346. https://doi.org/10.1177/1468797616680852.

Brennan-Horley, C., Luckman, S., Gibson, C., and Willoughby-Smith, J. (2010). Gis, ethnography, and cultural research: Putting maps back into ethnographic mapping. *Information Society*, 26(2): 92–103. https://doi.org/10.1080/01972240903562712.

Flyvbjerg, B. (2001). *Making Social Science Matter: Why Social Inquiry Fails and How it Can Succeed Again*. Cambridge: Cambridge University Press.

Flyvbjerg, B. (2004). Phronetic Planning Research: Theoretical and Methodological Reflections. *Planning Theory & Practice*, 5(3): 283–306.

Flyvbjerg, B. (2005). *Social Science That Matters*. Foresight Europe.

Flyvbjerg, B. (2006). Making Organization Research Matter: Power, Values, and Phronesis. In W. R. Clegg, S. R., Hardy, C., Lawrence, T. B. and Nord, W. R. (Ed.), *The SAGE Handbook of Organization Studies* (2nd ed., 370–387). Thousand Oaks, CA: SAGE Publications.

Gray, B. (1985). Conditions Facilitating Interorganizational Collaboration. *Human Relations*, 38(10): 911–936.

Gray, B. (1989). *Collaborating: Finding Common Ground for Multiparty Problems*. San Francisco, CA: Jossey-Bass Inc.

Hasse, J. C. and Milne, S. (2005). Participatory Approaches and Geographical Information Systems (PAGIS) in Tourism Planning. *Tourism Geographies*, 7(3): 272–289. https://doi.org/10.1080/14616680500164666.

Holm, L. K., Grenoble, L., and Virginia, R. (2011). A praxis for ethical research and scientific conduct in Greenland. *Études/Inuit/Studies*, 35(1–2): 187. https://doi.org/10.7202/1012841ar.

Hviid, P. and Beckstead, Z. (2008). Dialogues about Research. In M. Märtsin et al. (Ed.), *Dialogicality in Focus. Challenges to Theory, Method and Application*. New York: Nova Science Publishers.

Kesby, M. (2000). Participatory Diagramming: Deploying Qualitative Methods through an Action Research Epistemology. *Area*, 32(4): 423–435.

Kesby, M., Kindon, S., and Pain, R. (2005). "Participatory" approaches and diagramming techniques. In R. Flowerdew and D. Martin (Eds.), *Methods in Human Geography. A Guide for Students Doing a Research Project* (144–165).

Ladkin, A. and Bertramini, A. M. (2002). Collaborative tourism planning: A case study of Cusco, Peru. *Current Issues in Tourism*, 5(2): 71–93. https://doi.org/10.1080/13683500208667909.

Literat, I. (2013). "A pencil for your thoughts": Participatory drawing as a visual research method with children and youth. *International Journal of Qualitative Methods*, 12(1): 84–98. https://doi.org/10.1177/160940691301200143.

Marschall, A. (2013). Transforming subjectivity When aiming for mutually transformative processes in research with children. *Outlines: Critical Practice Studies*, 14(2): 160–183.

Marschall, A. (2017). When everyday life is double looped. Exploring children's (and parents') perspectives on post-divorce family life with two households. *Children and Society*, 31(5): 342–352. https://doi.org/10.1111/chso.12202.

Morris, J. C. and Miller-Stevens, K. (2016a). *Advancing Collaboration Theory: Models, Typologies, and Evidence* (Vol. 13). New York: Routledge.

Morris, J. C. and Miller-Stevens, K. (2016b). The State of Knowledge in Collaboration. In J. C. Morris and K. Miller-Stevens (Eds.), *Advancing Collaboration Theory: Models, Typologies, and Evidence* (3–13). Routledge.

Naalakkersuisut. (2015). *Turismeudvikling i Grønland – hvad skal der til? National sektorplan for turisme 2016–2020*. Retrieved from https://naalakkersuisut.gl/~/media/Nanoq/Files/Hearings/2015/Turismestrategi/Documents/Turismestrategi%202016-2020%20FINAL%20DK.pdf.

Pain, R. and Francis, P. (2003). Reflections on participatory research. *Area*, 35(1): 46–54. https://doi.org/10.1111/1475-4762.00109.

Ren, C. and Chimirri, D. (2017). *Turismeudvikling i Grønland – afdækning og inspiration*.

Ren, C. and Chimirri, D. (2018). Turisme i Grønland – før, nu og i morgen. *Tidsskriftet Grønland*, 4/2018: 295–302.

Ren, C., Jóhannesson, G. T., and Van der Duim, R. (2017). Co-Creating Tourism Research. In R. Ren, G. T. Jóhannesson, and R. Van der Duim (Eds), *Co-creating Tourism Research*. https://doi.org/10.4324/9781315393223.

Schram, S. (2012). Phronetic social science: An idea whose time has come. In B. Flyvbjerg, T. Landman, and S. Schram (Eds), *Real Social Science: Applied Phronesis* (15–26). Cambridge: Cambridge University Press.

Statistics Greenland. (2018). Greenland in Figures 2018. Retrieved from http://www.stat.gl/publ/en/GF/2018/pdf/Greenland in Figures 2018.Pdf.

Thomas, G. (2012). Phronetic social research: Putting power in context. Panel paper: "Research in context: Three approaches to thinking phronetically." Paper presented at the SPA/EASP York 2012 "Social policy in an unequal world" (July 17, 2012). York, UK.

Wood, D. J. and Gray, B. (1991). Toward a comprehensive theory of collaboration. *The Journal of Applied Behavioral Science*, 27(2): 139–162. https://doi.org/10.1177/0021886391272001.

9   Developing jurisprudence research through the engagement of students

Louise Faber

Background

Society needs lawyers, which creates an educational need. Educational needs create a need for teaching and jurisprudential research. Jurisprudential research is seen as an important element in the interaction that is a part of "the separation of powers." In short, the division of power implies the separation of the legislature authority, the judiciary authority, and the executive authority. For the legislative authority to be able to ensure that legislation is appropriate, legal research is necessary to systematize the decisions made by the judicial authority and derive the applicable law. The judicial authority also makes decisions as to whether the activities of the executive authority are legally enforceable. The executive authority must therefore also be able to ensure that legislation is appropriate. This is the second reason that research is necessary to systematize the decisions made by the judicial authority and derive the applicable law.

Jurisprudential research is, thus, essential to democracy, and the separation of powers can only work if there is a court and parties capable of producing the content of a case in the legal context so that a judge can make a decision. Lawyers and the education of lawyers are therefore also essential in a democratic country.

The lawyers currently serving in Greenland are predominantly Danes who have either migrated or work there on a fly-in/fly-out basis. The number of Greenlanders with a law degree is limited. This is problematic from a Greenlandic perspective, as it challenges the legal security of Greenlanders.[1] Similar challenges are faced in the Faroe Islands.[2]

An investigation of the possible causes found that it was problematic that the Greenlandic youth who aspired to pursue legal careers were able to fulfill these ambitions only by completing a law degree in Denmark. The Danish legal education is mainly taught in Danish and is obviously primarily about Danish law.

Moreover, a preliminary study revealed that Greenlandic students in Denmark struggle with the linguistic requirements of the courses.[3] They were suffering from cultural difficulties and experiencing a lack of motivation and homesickness. Such issues can complicate the completion of any course, but they especially complicate the legal courses, where the exams require a high degree of language proficiency.

The advisory board at Aalborg University, which includes Danish representatives from the industry, communal- and government authorities, the legal profession and the courts of law, insisted that Aalborg University should help to mitigate this problem.[4]

Considering the physical location of Aalborg University in North Jutland, where both authorities and enterprises traditionally have collaborated closely with the authorities and enterprises in Greenland, a decision was made to develop a legal program at Aalborg University in cooperation with the University of Greenland aimed at increasing the number of young Greenlandic lawyers.[5] Hence, the idea for the "North Atlantic Law Programme" (NALP) was created.[6]

The author of this contribution led the program in the formative phase (2011–17). This chapter therefore also contains a status of the experiences and fulfillment of objectives throughout this six-year period.

The fundamental idea in the short term

The project was intended to increase the number of young people with a Greenlandic background who obtain a law degree. This objective included four subordinate objectives:

1   Increase the admission rate of applicants with a Greenlandic background
2   Reduce the dropout rate of admitted Greenlandic students
3   Establish education options in Greenlandic law
4   Create both the basis and motivation to use these acquired capabilities to contribute to a further accumulation of knowledge in Greenland upon graduation

The third and fourth objectives are particularly relevant to this chapter; namely, the general objective of NALP included an ambition that, upon graduation, the students should possess the competence and drive to function as lawyers, judges, etc. in the legal system in Greenland, thereby contributing to the development of the sector. As it is assumed that the students have a Greenlandic background, the development will be based on an understanding of Greenlandic culture, values, and traditions to a greater extent than previously. In so doing, even a small educational initiative can contribute to capacity-building in Greenland.

The collaboration with the sector

The fundamental idea became to begin collaborating with the authorities on the Greenlandic legal system (the sector) on the development of options and professionalism for students. It quickly became apparent that it was difficult to provide educational material regarding subjects concerning Greenlandic law without cooperating closely with the sector.

This cooperation was necessary, owing to insufficient access to the sources of law, which is crucial to research using legal method and logic.[7]

The most important sources of the law include the legislation decided by the self-government in Greenland. The legislation must be interpreted in the light of the legislators' remarks on the law, not all of which were accessible online.[8] The same applied to the interpretive notes.

The sources of law include the verdicts made by the courts of law in Greenland and the Supreme Court of Denmark. In Denmark, judicial decisions are systematized in journals and made publicly available. Passed verdicts made by the Greenlandic courts were not readily publicly available at the time and could be acquired only via a request for access to documents in the internal court archives, either by inquiring with the lawyer's offices or by addressing the prosecuting authority. There were no systematic listings or archives accessible to the public, nor were there any online search engines for acquiring the aforementioned verdicts. Those verdicts that could be procured were thus acquired rather randomly. The lack of systematic access to sources of law renders the production of jurisprudential contributions about Greenlandic law difficult.[9]

Collaborating with the sector was also rendered necessary by the lack of literature; few scientific contributions about Greenlandic law existed. Knowledge of this could therefore be obtained only by applying the experiences of others, particularly experienced practitioners in the sector.

Moreover, most of the acting lawyers in the (broadly defined) legal system (the courts, lawyer's offices, prosecuting authorities, and other authorities) had Danish roots without any Greenlandic background.

Many of the lawyers were also working as part of a relatively brief stay to advance their personal careers. The knowledge acquired during such a brief stay is usually lost in that it is neither embedded in the society nor in the legal system and is rarely conveyed in any way subsequent to the termination of employment.

The reasons mentioned above are important because they are interconnected and have an amplifying effect on each other. They pose a problem that can be solved only if the legal positions are disputed by Greenlanders.

The Greenlandic and Danish graduates of these law programs must acquire knowledge in Greenlandic law upon graduation. This is complicated by the insufficient access to the sources of law and the systematic listings and propagations in the form of jurisprudential contributions hereof. Consequently, the collection of sources of law and reprocessing of knowledge and skills in Greenlandic law are only really possible through one's legal experience. This need will not be met without sufficient numbers of Greenlandic lawyers and vacant positions repeatedly being occupied by short-term Danish lawyers.

The development process needed to create access to the sources of law and jurisprudential research is therefore complicated by the short-term employment of Danish lawyers. But such short-term employment is made necessary by the lack of access to the sources of law and the ever-absent Greenlandic lawyers.

These circumstances are a significant hindrance to maintaining legal protection, as the knowledge of applicable law is a fundamental premise for a well-functioning society, as mentioned above. It is a difficult circle to break. In practice, the need for labor overshadows the long-term need to procure and propagate knowledge of and scientific contributions to Greenlandic law.

At the same time, the insight described above confirms the need for a focus on improving the education of Greenlandic lawyers, both in quantity and quality.

Program content

Trying to find a concrete solution to the issues relating to the missing educational material, Aalborg University decided that the students needed to acquire the necessary skills by making a problem-based project in cooperation with the sector in Greenland. The Greenlandic students were therefore offered a project-oriented course in Greenland in the final semester of the bachelor program. The offer was implemented as a supplement to the curriculum.[10] The course was provided as an alternative to an elective subject and the so-called "bachelor project," which would typically be completed in Denmark. The project-oriented course offered to the Greenlandic students corresponds to two-thirds of a semester.[11]

To support the cooperation with the sector in Greenland, an independent advisory board with Greenlandic representatives experienced in specific sectors was established to support the development of the NALP.[12]

By now we lacked only the students. The admissions process for Greenlandic students at Aalborg University includes a partnership with The Greenlandic House in Aalborg, which was therefore an obvious first step toward getting in touch with the students.

Person-to-person training: Guided study group

The students are often initially in contact with The Greenlandic House. After beginning their studies, newly admitted Greenlandic law students are promptly invited to a meeting with other participants, including guidance counselors from The Greenlandic House. Not all of the students have heard of the NALP, and the meeting is held to create the basis for a partnership between new students and other NALP participants. Here, the intention is to ensure personal and academic development by offering a study group guided by a legal researcher. I undertook this task as the former head of the NALP.

The guided study group is held on a monthly basis and is used to discuss shared challenges and solutions to the aforementioned challenges, namely to establish tutor- and mentor coaching and to plan courses and exam preparations. The guided study groups are also used to discuss bachelor projects, which are planned well in advance to ensure that each student has the optimal opportunity to prepare their subject choice and sector-cooperation in Greenland. In addition, help is needed to prepare application for funds to cover travel costs and their housing during the stay.

These practical subjects comprise much of the guided Study Group, although it does also have a pedagogical purpose. The Greenlandic students differ from the Danish students. They have completely different backgrounds and different expectations for the future. They often find it more difficult to see the purpose of their education and are less optimistic about their future prospects than are Danish students. This has a negative impact on their engagement, not least once they learn that the legal course is more difficult to complete than they anticipated. It is important that the students receive assistance to deal with these problems to keep the course-completion percentage at a reasonable level. The study group is therefore also a forum where positive expectations, engagement, identity, and resources for handling setbacks can be established.

In this respect, meeting older students with similar backgrounds who share experiences and encouragement is a source of motivation for the younger students. The regular meetings with dedicated educators and guides who are equally committed to developing the student personally and academically contribute to the same end. The student–guide partnership creates a safety net that makes it easier to handle setbacks and to keep faith in the future.

One important pedagogical focal point is to help the student to understand that low grades and even failed exams are to be expected. It is necessary to accept this as a premise of the first years of the program, owing to the cultural differences and difficulties with the Danish language, as it takes years to level out these differences. The study group also nourishes the faith in being able to complete the program, despite setbacks and faith in being able to find employment, despite what appears to be a low academic level.

Working with the students to accept their grades obviously requires insight and understanding of the sector in which the students want to gain employment. This is a challenge for communications, but it is also helped by the cooperation with the sector, where it is possible to convey the competencies obtained by the students during their education without referring to grades.

Contents of the project course

The core purpose of the project-based coursed consists of the demand of physical residence in Greenland with connection to an authority or business.

The course is held over a 2–3 month period, often in February–May, and the students are obligated to find a partner in the sector with whom they can cooperate. This partnership helps the student to define suitable legal issues of relevance to Greenland and renders it possible to obtain access to the sources of law needed to work with these issues. The partnership also makes it possible for students to build and maintain a physical and academic connection with their partner in Greenland. Notably, the project-based course is the initiative that largely necessitates creating a partnership with the legal sector in Greenland.

Demands pertaining to the academic content of the project

The curriculum defines the required project outcome as giving the student knowledge of Greenlandic law and legal issues associated with the use of legal logic in Greenlandic law. This includes finding and choosing legal sources of relevance to the legal issue about which the student is intending to write.

As mentioned above, however, the sources of Greenlandic law are difficult to access, meaning that students must invest great effort in acquiring usable case law. In the evaluation of the academic quality of the project, it is therefore necessary to modify the demands in the curriculum pertaining to the student's use of legal sources. Thus, it is accepted that the student's presentation concerning Greenlandic Law relating to a project-based course only contains a repetition of the law, available preliminary work, and some acquired verdicts, even if they provide only an illustrative image of the applicable law.

These modified demands can be justified, as the skills that the student is supposed to acquire are defined in the curriculum as to be able to use legal terminology and expressions, to express oneself clearly and with structure regarding the issues that relate to Greenlandic law by independently and competently using all necessary sources of law in the argumentation.[13] As shown, there are no demands stating that the student must be able to determine what applicable law is. Nevertheless, the student is expected to use the legal logic to conclude what might be the applicable law.

The student must also follow the instructions in the curriculum regarding the completion of bachelor projects in general, including the description of their practical and methodical difficulties in acquiring sources of law for the project. This is to include mention of whether the project contains all sources of law or (more likely) if it contains only random verdicts and the like. The student must therefore also consider the impact that this has on their conclusion concerning applicable law.

Because the sources of law are also unavailable to the examiner and external examiner, an appendix to the project must include any legal sources that are not publicly available, as well as a concise description of the organization or authority with which the project was created.

Partnerships and the projects

From 2015 until late 2018 project-based courses were completed in collaboration with Nuna Lawyers (Nuna Advokater) and the Meinel legal firm (Advokatfirmaet Meinel). Both of these companies have offices in Nuuk, where the students have been established.[14] The Court of Law in Greenland, The National Court in Nuuk, and some of the district courts have all participated in the collaboration as project partners or by giving the students access to sources of law and helping to answer various relevant questions.[15] The privately operated Collection of Greenlandic Law has granted payment-free access to the students. Other public authorities, especially the Parliamentary Commissioner, have provided charitable assistance. Ilisimatusarfik (the University of Greenland) has provided study facilities, including access to the library and internet. Ilisimatusarfik has also assisted in the examination of students physically located in Nuuk. Part of the NALP program is for students to be able to take both written and oral exams (for example, using Skype) in Nuuk. This is also an option when completing a bachelor project in cooperation with a partner in the sector in Greenland.

Alone or in groups?

The bachelor project can be completed both individually and in groups of as many as three people, although most of the projects thus far have been written individually. The limited number of Greenlandic students in each annual cohort renders it rarely possible to complete the project as a group. One project was completed by a group consisting of one Danish and one Greenlandic student in 2016, and another was completed by a group consisting of two Greenlandic students in 2017.[16]

The exam and achieved competencies

Both the project and the exam are intended to develop the students' competences in their use of subject-related sources of law and judicial methods in terms of solving a judicial issue in Greenlandic law. In addition, the course develops competencies in expressing common scientific questions and reasoning in a clear and concise manner and in a structured, legal language. Such a skillset is necessary to be able to write jurisprudential contributions about both Danish and Greenlandic law.

The exam is held as a regular bachelor project exam in the law degree program, meaning that the exam is concluded by an individual oral examination with an external examiner based on the project. This can be held in person at Aalborg University or via Skype if the student is physically at the University of Greenland. Thus far, the students have preferred to take this exam in person in Denmark. Only one exam was held with the assistance of the University of Greenland, but the students participate gladly in other examinations with Ilisimatusarfik's assistance.

Grading in the exam is based on the project, but other curriculum-related questions relevant to the project can be asked, including questions about Danish law. The grading also includes an assessment of the oral presentation itself, exactly as any other bachelor project delivered as part of the law degree program. Thus, the grading of the academic quality of the project and the Greenlandic student presentations are based on the same standards as any other bachelor project in the law program. Difficulties experienced in acquiring the sources of Greenlandic law are the only special consideration in the assessment. This is because the project must be graded as if the listings of the sources of law are comprehensive, even though they cannot be so, owing to the circumstances of their project.

The first bachelor project on Greenlandic law was delivered in 2015. It discussed a subject within family and parental custodial law and was awarded the highest possible grade, an A.[17] In 2016 a student wrote a project about how and when a licensed defendant is appointed in Greenland according to the Act of Greenlandic Administration of Justice.[18] Another student wrote about the rule of law in the District Courts of Greenland. The latter project also received the highest possible grade, together with a commendation from the Danish Bar and Law Society.[19] In 2017 a two-person group completed a bachelor project on criminal law, and both students were awarded an A.[20]

The students participating in this exam have all fared well. In fact, they have done even better on this exam than they have on the other exams that are part of their bachelor program. This is partly understood as owing to the Greenlandic students not having to deal with the language-related difficulties that had affected their earlier exams.[21] But the better performance is also the result of the Greenlandic students being able to draw their own conclusions based on an analysis of the sources of law using legal logic. It is namely the student's analysis of the sources of law that is the subject examined in both the bachelor project exam and the examination of the candidate thesis.[22] The Greenlandic students' projects are thus presenting the purest form of forensic craftmanship without explanations and summaries of the works of other authors, which can unintentionally disrupt the use of legal logic in projects and theses about Danish law.

In conjunction with the writing of this contribution, the NALP students have been asked if they would recommend writing a project about Greenlandic law in cooperation with a Greenlandic law firm or public authority. The students generally agree on the project being an educational and motivating experience. One student insists that it is "the best way to learn the theories. It gives so much motivation; you start getting excited about being a part of the job market and a tiny 'academic seed' is planted, where you start considering the different problems you could work with."

The thesis

The candidate course is concluded with a 30-ECTS (European Credit Transfer System) thesis. An entire semester is therefore set aside to write the thesis, and students have the option to write about Greenlandic law. If the student is able to find their own housing, the thesis can be completed in Greenland. Although it is not compulsory, the students also have the option of completing their thesis working in cooperation with a public authority or company in Greenland.

Three students had completed their theses at the time of writing. The first thesis about Greenlandic law was delivered in 2016. It dealt with a subject in family and inheritance law and was awarded an A.[23] No Greenlandic candidates completed the law program in 2017, but two candidates finished in 2018—one with a thesis about criminal law and the other with at thesis about computer fraud. All three students have partnered with a law firm and the Court in Greenland to have a physical workplace and the opportunity to discuss and access sources of law.

Achievements

In total, four bachelor projects about Greenlandic Law were completed by five Greenlandic students through this arrangement in the period 2015–18. This means that there are now five Greenlandic students with bachelor-level competencies in Greenlandic law. Three of these students have also graduated with a master's degree, two of whom have gained employment as lawyers in Greenland. The last of the three graduates wishes to pursue employment in Greenland later if the opportunity presents itself but is still residing in Denmark for family reasons. Henceforth, one or two candidates will hopefully graduate every year with an insight into Greenlandic law, assuming that the NALP will continue. This is an important improvement compared with the situation before the initiative began.

In 2013 the program was endorsed by the Joint Committee, which is a government coalition between Greenland, USA, and Denmark that was founded to create and develop academic, cultural, and professional bonds between said countries. This includes the sharing of research, scholarships, and student-to-student activities. The joint committee grants prestigious endorsements to active projects containing a declaration that said project benefits the Joint Committee's cause. The endorsement does not give any specific entitlements on its own, but it is created as a document that can be attached by the project management, employees, or students seeking financial support for a project or housing. The document increases the probability of receiving resources from such applications inasmuch as it documents the societal relevance of the project.

In 2015 the program was mentioned as a "best practice case" in a report made for the Nordic Council of Ministers, which sought to investigate new ways of securing the education of young people in outlying Nordic areas.[24]

The attention that the initiative has drawn to the unique difficulties and needs of Greenlandic students has helped to contribute to improving their situation. A practical detail that also boosts motivation is that relevant student jobs are becoming available that benefit Greenlandic students, including in Aalborg.[25] In Greenland, summer jobs for students have been established. These jobs are typically with the public authorities and companies that have acted as project partners, but other bodies have also become involved.

The NALP has been a contributing factor in the University of Greenland managing to gain political support and financing to create a bachelor's degree program entitled "Jurisprudence and Public Law" in 2018. The Greenlandic law program teaches its students skills for academic positions or work as a consultant in the Greenland Home Rule, in a municipality, or in a publicly owned company. If the student wishes to pursue graduate studies in law or to become a lawyer working in the legal sector, they must still pursue a law degree in Denmark; probably with some transfer of credit for the subjects that they have completed about public law.

The long-term purpose of the NALP was to contribute to the further accumulation of (legal) knowledge in Greenland. The program has demonstrated that this is possible. The long-term purpose was also to develop jurisprudence research by better engaging students. Although the program has yet to prove this to be possible, the students have produced projects and theses on Greenlandic law, the results of which have yet to be made public. This is the next goal of the program. Hopefully, the students' results will soon be published in a journal for this purpose. The journal could also provide other options for publishing articles on Greenlandic law. This will lay the foundation for further developing subjects and teaching materials and, in the long term, the publication of actual jurisprudence research.

For the duration of the course and through the study group, the Greenlandic legal students also developed a study community and developed a sense of solidarity. After graduation, they therefore have a basis for continued academic networking in Greenland. The Greenlandic students have already established a Facebook-based alumni network. This goal is attempted realized by continuing to guide the alumni with a jurisprudential approach. Through continued guidance, it may be possible to motivate candidates for enrolment as PhD students.

Even though the NALP is a fairly simple initiative with few resources, the engagement of the students has had a significant impact on the sector and the desire to produce legal research investigating Greenlandic law. In the long term, this might lead to a research environment and the development of jurisprudence research.

Notes

1 This concern is very current. See, e.g., Kim Rosenkilde: Aaja Chemnitz: Mangel på jurister skader retssikkerhed i Grønland [The lack of lawyers hurts the rule of law in Greenland], Altinget, January 29, 2017, www.altinget.dk.

2 Around this period, Professor Kári a Rogvi also took the first steps towards establishing a law program in the Faroe Islands.

3 This is a general problem. See, e.g., Heidi Kokborg, Kristine Buske Nielsen, and Thomas Christoffer Bahr: Den snørklede vej til en dansk uddannelse [The

meandering path to a Danish education]. For reference, see: http://groenlandskes tuderende.mediajungle.dk/?fbclid=IwAR20lRwpLaUz-iJz4ob1ExWxH3L_Fm Z3NacaDLouZ41SukIAJPzBvd_MZLY

4 The University of Southern Denmark simultaneously established a distance education program in Danish law that also targets Greenland residents. See: https://www.sdu.dk/da/om_sdu/institutter_centre/juridisk+institut/nyheder/fondspenge.

5 Aalborg University and the University of Greenland entered into a 5-year cooperation agreement for the period March 2012–March 2017. The terms of the agreement can be found at www.aau.dk/om-aau/internationalt-samarbejde/groenla nds-universitet

6 See the last page of the Strategy of the Law School 2012 to 2015. The strategy is not publicly available but can be obtained upon request from the law program at Aalborg University.

7 As in Denmark, The Greenlandic legal system is based on Civil Law. The scientific procedure used in legal systems based on Civil Law, referred to as judicial method, is also commonly referred to as legal logic.

8 The legislation adopted since 2010 can be found at www.lovgivning.gl. Some of the older legislation is accessible at www.arkiv.lovgivning.gl. The legislation can also be found in the Greenlandic Law Collection, a privately run law collection, at www.dgl.gl.

9 This is also criticized by The Lawyer Council (Advokatrådet) in the report on legal certainty in Greenland: Retssikkerhed i Grønland (2016): p. 13.

10 See point 8 of the "Study Board meeting on 31 January 2014." The record is publicly available at https://www.law.aau.dk/digitalAssets/90/90034_referat-13—janua r-2014—hjemmesiden.pdf. See also the supplement to the curriculum, "North Atlantic Law Program, Grønland, Tillæg til Studieordning Den juridiske bacheloruddannelse JURA, Aalborg Universitet, Gældende fra September 2014," https://www.fak.samf.aau.dk/digitalAssets/94/94767_jurba-nalp14.pdf som. The project-oriented course is described in section 12a.

11 The last ten ECTS in the sixth semester are made up of a compulsory subject. The subject will be admitted to the program in light of the individual student's specific course of study before or after the bachelor project. The students can also complete this subject on their own during their stay in Greenland.

12 An advisory board with representatives with knowledge of the Greenlandic sector was appointed by the Dean of the Faculty of Social Sciences, Hanne Kathrine Krogstrup, on January 23, 2014, and existed until 2017. The representatives were Klaus Georg Hansen, The University of Greenland; Jan Thryesøe (later Sørens Stach Nielsen), The Greenlandic House; Thorkild Mørk Rønbøl Lauridsen, Arctic Consensus; Lise Lotte Terp, Aalborg Harbor and Logistic A/S and Arctic Consensus; Thomas Trier Hansen, Nordic Law Group ApS; Christian Lundblad, The Court in Aalborg; Anders Hjulmand, the HjulmandKaptain legal firm; Per Vestergaard, the LETT legal firm; Evan Christensen, the auditing company EY; and Rikke Albrektsen, Frederikshavn Municipality. The NALP Advisory Board was combined in a small version with the Law Board "General" Advisory Board in 2017.

13 See the supplement to the curriculum, "North Atlantic Law Program, Grønland, Tillæg til Studieordning Den juridiske bacheloruddannelse JURA, Aalborg Universitet, Gældende fra september 2014," https://www.fak.samf.aau.dk/digita-lAssets/94/94767_jurba-nalp14.pdf som. The project-oriented course is described in section 12a.

14 The law firm profiles can be found at nuna-law.com and meinel.gl.

15 Information about the Greenland court system can be found at gl.domstol.dk.

16 For references, see Jesper Hansen in Sermitsiaq: Grønlandsk-dansk juraprogram ruller [The Greenlandic-Danish legal program is rolling along], August 11, 2016, at www.sermitsiaq.ag and Jesper Hansen in Sermitsiaq: Flotte karakterer til

grønlandske jurastuderende [Good grades for Greenlandic law students], June 28, 2017 at www.sermitsiaq.ag.
17 "12" in the Danish grading system. See Jesper Hansen in Sermitsiaq: 12-tal til grønlandsk jurastuderende [Greenlandic law student gets an A], June 25, 2015, www.sermitsiaq.ag.
18 See Jesper Hansen in Sermitsiaq: Grønlandsk-dansk juraprogram ruller, August 11, 2016, www.sermitsiaq.ag.
19 See Jesper Hansen in Sermitsiaq: Grønlandsk-dansk juraprogram ruller, August 11, 2016, www.sermitsiaq.ag.
20 See Jesper Hansen in Sermitsiaq: Flotte karakterer til grønlandske jurastuderende, June 28, 2017, www.sermitsiaq.ag.
21 See Jesper Hansen in Sermitsiaq: Flotte karakterer til grønlandske jurastuderende, June 28, 2017, www.sermitsiaq.ag.
22 See Jesper Hansen in Sermitsiaq: Flotte karakterer til grønlandske jurastuderende, June 28, 2017, www.sermitsiaq.ag.
23 See Jesper Hansen in Sermitsiaq: Flotte karakterer til grønlandske jurastuderende, June 28, 2017, www.sermitsiaq.ag.
24 See Stine Thidemann Faber, Helene Pristed Nielsen, and Kathrine Bjerg Bennike, *Place, (In)Equality and Gender: A Mapping of Challenges and Best Practices in Relation to Gender, Education and Population Flows in Nordic Peripheral Areas* (Nordic Council of Ministers, 2015): 141–43.
25 A position as interpreter was established in the Job and Integration House in Denmark, for example.

References

Aalborg University. (2012). Collaboration Agreement between Aalborg University and the University of Greenland for the period March 2012–2017. Retrieved fromwww.aau.dk/om-aau/internationalt-samarbejde/groenlands-universitet.

Aalborg University. (2014). North Atlantic Law Program, Grønland, Tillæg til Studieordning Den juridiske bacheloruddannelse JURA, Aalborg Universitet. Retrieved fromwww.fak.samf.aau.dk/digitalAssets/94/94767_jurba-nalp14.pdf.

Faber, S. T., H. P. Nielsen, and K. B. Bennike. (2015). *Place, (In)equality and Gender. A Mapping of Challenges and Best Practices in Relation to Gender, Education and Population Flows in Nordic Peripheral Areas.* Nordic Council of Ministers: 141–143.

Grønlands Domstole. (2020). Om Grønlands Domstole [About the Greenland Court system]. Retrieved fromwww.gl.domstol.dk.

Hansen, J. (2015) 12-tal til grønlandsk jurastuderende [Greenlandic law student gets an A], *Sermitsiaq*, June 25. Retrieved fromwww.sermitsiaq.ag.

Hansen, J. (2016). Grønlandsk-dansk juraprogram ruller [The Greenlandic-Danish legal program is rolling along], *Sermitsiaq*, August 11. Retrieved fromwww.sermitsiaq.ag.

Hansen, J. (2017). Flotte karakterer til grønlandske jurastuderende [Good grades for Greenlandic law students], *Sermitsiaq*, June 28. Retrieved fromwww.sermitsiaq.ag.

Kokborg, H., Nielsen K. B., and Bahr, T. C.Den snørklede vej til en dansk uddannelse [The meandering path to a Danish education]. Retrieved fromhttp://groenlandskestuderende.mediajungle.dk/2017/11/21/hello-world/.

Rosenkilde, K. and Chemnitz, A.(2017). Mangel på jurister skader retssikkerhed i Grønland [The lack of lawyers hurts the rule of law in Greenland], *Altinget*, January 29. Retrieved fromwww.altinget.dk.

10 Recruiting and retaining labour in Greenland

A PhD project in close cooperation with local stakeholders

Verena Huppert

Researchers can find cooperation on research projects with local stakeholders both advantageous and challenging. In Greenland, collaboration with locals in research projects has become increasingly common in recent years, in contrast to the extensive "hit-and-run"[1] research that has gone on for many decades. On their website, the Greenland Research Council states that their main aim is "to ensure the best possible knowledge base for the development of Greenlandic society" and to assist in the creation of knowledge for Greenlandic society (Greenland Research Council, n.d.). By this, they do not directly criticize hit-and-run research, although they emphasize the importance and indirectly support the creation of more research projects in collaboration with the local community.

The point of departure for this chapter is my ongoing PhD project, the working title of which is, "Recruiting and retaining labor in Greenland in the light of the skills gap," which is scheduled for completion in mid-2020. The main purpose of the PhD study is to examine the labor recruitment and retention challenge in Greenland, on a labor market characterized by a skills gap. By studying different employee groups and investigating how the employees and employers alike perceive the challenge and what the employers are doing about it, it is aimed at identifying gaps between perceptions of the challenge and existing structures. I started working on the project in April 2017.

The project is co-financed by the Department of Culture and Global Studies at Aalborg University, Arctic Consensus,[2] the Sermersooq Municipality, and the Greenland Business Association. From the outset, the involvement and active participation of local stakeholders has been regarded as crucial to anchor the project in the Greenlandic context. This necessity evolved both from the researcher's own upbringing in the tradition of research that must be beneficial for society, as well as this particular topic of research. While this type of research is not beneficial for all topics, the research topic in this case presents an issue that makes activating and collaborative research necessary.

Most of the period of research was to be spent in Greenland, the intention of which was to produce deeper insight into the field of research. In the summer of 2017, I formally moved to Nuuk, returning to Aalborg University roughly twice a semester.

The present project is engaging in participatory research (PR) involving local institutions as stakeholders together with local partners. However, the level of involvement of these partners is slightly different than in traditional PR. Traditionally, PR is characterized as community-based research, where the scientist collaborates with a community to investigate their realities. The members of the community are to be the main beneficiaries of the research. The problem to be investigated is defined based on the community's own realities, and it is analyzed and solved in collaboration with the community. PR requires the community's full and active participation throughout the research process (Guthrie, 2010; Hall, 1999). The community under investigation in this project consists of the employers and employees in Greenland. In order to collaborate with this community, selected employer and employee groups were chosen for the study. The problem formulation and research questions have been designed in close collaboration with the selected groups and local partners, who have been in close contact with the researcher during all stages of the study. While the researcher steers the project and conducts the data collection, the partners and studied groups have been continuously involved in discussions on which data to collect and how to interpret and analyze this data. This means that even though there is no active involvement in the data collection (apart from the community being the main informants), there is still a high level of involvement toward achieving the final learning process, which is one of the main goals of PR. In addition, the topic of the study is of utmost relevance for the involved partners, which also explains their interest in contributing financially to the project. I argue that even though the partners' monetary contribution was of crucial importance for establishing the project, the study can be seen as participatory, community-based research, as it involves a community actively throughout the research process with the ultimate goal of benefiting the community under investigation. The fact that the community partly consists of paying partners makes no difference to the value and justification of this project. Local relations and a local network were necessary to investigate the particular research question and probably would have been the same, even if the project had been exclusively financed by a university.

The chapter begins with a section on how the project came to be. From that follows a section on the research design and methodology, which presents its pragmatic and explorative design and how the collaboration with partners works in practice, followed by discussion of the practical experiences. As this is an ongoing research project, the chapter focuses primarily on the process before and during the research project rather than on the outcomes. It outlines the potentials of research-based partnerships with local stakeholders and own experiences with PR.

How the project came to be

On a practical level, this PhD project involves several partner types. Some partners were involved in the creation of the project and support it financially and/or administratively (referred to in the following as "Category 1 partners"), while others serve as cases for the study (referred to in the following as "Category 2 partners"). Some partners belong to both categories, as they both provide support for the project and data is collected from some of their employees. In total, this project collaborates with five non-academic partners. In addition to the official partners, I was contacted by local informal informants over the course of the project.

The project came into place in extension of a master's degree thesis at Aalborg University on the retention of local labor in Nuuk, Greenland, and Charlottetown, Prince Edward Island, Canada (Huppert, 2016).

Several factors contributed to the project, which was enhanced by experience from previous work in the field:

- An extended, unique network with relevant local and non-local partners
- Extended understanding of the importance of the project and the local collaboration
- Unique insights into the topic and the potential of further research for both academia and society
- A basis of trust between partners and the researcher

The thesis project was carried out in cooperation with Arctic Consensus, the Sermersooq Municipality, and the Greenland Business Association. During the master's thesis period, fieldwork was conducted in Nuuk and Charlottetown, and good contact with the collaboration partners was established. This phase served as a steppingstone for establishing the PhD project in that it raised awareness of the above points and contributed to the idea for how to form a PhD project with the strong participation of local partners with whom contact had already been established. Lofland and Lofland (1984) argue that one should "start [qualitative research] where you are − to use your current situation or past involvement as a topic of research" (Lofland & Lofland, 1984: 2; in

Figure 10.1 Collaboration continuum
illustrates the placement of the partners on a "collaboration continuum," as they can fill different roles at different times, including stakeholders, cases, informants, and informal informants.

Wadel, 1991: 29). In a sense, this reflects the creation of this project, as it slowly revealed the value of local collaboration in such research projects.

The research for the master's thesis provided a head-start on a PhD project on an associated topic, as the scope of the topic and the relevance for further research had already been made clear. Throughout the fieldwork in Nuuk, the importance of the topic for the local community became clear, including the local stakeholders' interest in it and the potential for further cooperation. After finishing the thesis, its reception and the persistent inquiries about the thesis from local stakeholders illustrated the potential for further work in the area.[3] This demonstrates how an early or previous engagement with the field can have a positive impact on future collaborations.

For almost two years up to the official beginning of the project, extensive effort was invested in personal networking with stakeholders both in Greenland and Denmark. Being a cooperative project between a university, two public stakeholders, and a privately financed institution, the creation of the PhD project was based on the relations developed with the involved institutions, and in particular their trust in the researcher and in the project. Trust can be argued as being fundamental in research, especially in PR. It cannot be bought—only developed over time—which is why "hit-and-run" research rarely achieves trust. All of the Category 1 partners had worked with me before—they knew me, my background, how I work, and they understood my interest in the topic. Their support and investment in the project were not only based on the topic and the need for research in this field, but also to a high degree on the relationship and trust that had been built.

As shown in Figure 10.2 below, Category 1 partners were involved in the project from the beginning, as the project was established in cooperation with them. The contact to the Category 2 partners was first established after the project had started. The partners were chosen based on their relevance as resources for the study.

Instead of contacting different managers in the company directly, the focus was on anchoring the project in main management in order to base the project as an internal project, which was expected (and ultimately proved) to ease the

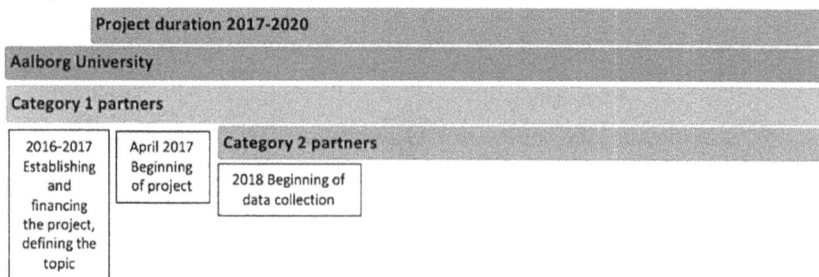

Figure 10.2 Involvement of partners during project duration

contact with the employees. Potential problems with this top-down approach were overcome by carefully choosing the employee groups in close collaboration with the respective leader and presenting a detailed explanation of the purpose of the project. The top-down approach was also beneficial for the fact that it would not have been possible to get in contact with the local managers without help from the management, for practical and geographical reasons.

The example of how this project came into being shows that one cannot "just" go out and conduct participatory research, as it is necessary to build a base, understanding, and trust. Network and relations are of exceptional importance for PR.

Like many other PhD projects, the establishment of this project (and not least the financing) took some time. A formalized partnership was established in early 2017, and the work on the project officially started in April 2017. The whole process—from pitching the concrete idea for this project for the first time to signing the contract—took about 1.5–2 years in total.

Research design and overall methodological approach

The research design and methodology for this study reflect the aim of grasping the realities of the studied context (i.e., the Greenlandic labor market). It combines pragmatism, applied phronetics, and an explorative design with mixed methods. The choices are informed by the research question and the underpinned objectives that focus on the multi-dimensionality of the realities studied. The research design is informed by the project's aim to derive knowledge from practice and to create an outcome that adds to the capacity of society.

This type of research is conceived as the means to create practically applicable knowledge. The fact that this research has external, non-academic funders is closely connected to the need to create applicable knowledge for the benefit of the involved stakeholders; in this case, employers and employees in Greenland. The project has involved local partners from the beginning and plans to continue doing so throughout project, joint decisions especially being made early in the project, such as specific focal points, albeit with a strong focus on the researcher being in charge of deciding the overall direction of the project.

The philosophical point of departure for the study is based on a pragmatic approach to the complex realities encountered during the course of the study connected to the approach of applied phronetics (Flyvbjerg, 2001: 60ff).

The topic and general angle of the project were chosen based on its relevance for academia and society alike. Based on meetings held in the autumn of 2016, the initial project idea was presented and discussed with Category 1 partners, and elements of their feedback were included in the project outline. The project was presented to the Category 2 partners in the first months after starting it, and they were asked if they might be willing to contribute to data collection as cases. They had no influence on the general angle of topic.

Morgan (2014) claims that pragmatic research must be concerned both with the *"how to"* and *"why to"* do research:

> When we ask "why to" questions, this points to the importance of our choice of research goals. Yet even the "how to" questions involve more than making technical decisions about research methods because of the commitments we make when we chose one way rather than another to pursue our goals.
>
> (Morgan, 2014:1046)

This project's *"why to"* is concerned both with its relevance for society and the stakeholders. The research goals were chosen to some degree in collaboration with them or in accordance with some of the observed challenges facing the stakeholders. Further, in the context of this project, the *"why to"* is strongly connected to the *"how to"* do research. Constructing the research question around the challenges experienced by local stakeholders renders the research relevant by definition. This relevance is further secured by the joint formulation of the research question. In addition, extended access to data is enhanced and the credibility of the data is further ensured, as the partners are continuously involved in the project. Obviously, the practical knowledge of the field possessed by the local stakeholders is of great value to the project.

Incorporated in the pragmatic and explorative approach of this study, the following statement by Flyvbjerg (2001) matches this research project's approach to collaboration with local partners while at the same time maintaining a critical distance:

> One gets close to the phenomenon or group whom one studies during data collection, and remains close during the phases of data analysis, feedback and publication of results ... this strategy typically creates interest by outside parties, and even outside stakeholders, in the research. These parties will test and evaluate the research in various ways. The researchers will consciously expose themselves to reactions from their surroundings − both positive and negative − and may derive benefit from the learning effect, which is built into this strategy. In this way, the phronetic researcher becomes a part of the phenomenon studied, without necessarily "going native" or the project becoming simple action research.
>
> (Flyvbjerg, 2001: 210)

Again, the joint creation of the research question can be highlighted as assuring the relevance of the project. The method used in this study does not aim for me as researcher to conduct action research,[4] I myself becoming a subject in the study, but rather dialogue and interaction with the participating partners to a degree that respects a critical distance. The approach focuses on the learning process for the researcher and participants alike, as it uses reactions and informal informants to test and evaluate the research.

The pragmatic approach to this research project and the *"how to"* also relates to Kørnøv, Lyhne, Larsen, and Hansen's concept of a "change agent" (Kørnøv, Lyhne, Larsen, and Hansen, 2011). The authors argue for the relation between the researcher and external partners being characterized by interdependence and joint decisions on frames, while the researcher decides on the direction of the project (Kørnøv et al., 2011: 207). In this way, the researcher maintains their independence and critical distance at the same time that the project has the "potential to improve the connection between research and practice and promote sustainable development" (Kørnøv et al., 2011: 203). The research is based largely on the researcher–external partner relationship, which requires strong commitment from both sides. The fact that the research topic should be relevant for both academia and society in general might ease this cooperation. In the case of the project in hand, the topic has been an ongoing challenge for both the Category 1 and Category 2 partners, and the collaboration on the topic was experienced as rather smooth. The partners are interested in providing data and knowledge, as they hope that the research will shed new light on, raise further awareness regarding, and potentially produce solutions to an acknowledged but as yet unsolved problem.

The decision to move to Greenland shortly after starting the project is rooted in the pragmatic understanding of how to conduct research. Extended access to the field was required in order to gain a deeper understanding of it and the particular challenge that the research addresses. A researcher can never be entirely objective regarding their topic of research, and there is a strong awareness of how the researcher's background has an impact on their research. Subjectivity is not necessarily a bias, however, as it also defines the character of human understanding (Greene, 2007: 40); given extended access to the field, the researcher's understanding will evolve. Gibbons (1999) uses the terms *"contextualization"* and *"contextualized knowledge"* to describe how the influence of society can transform science (Gibbons, 1999: C82). The interaction with the local partners and informal informants is expected to transform the research project in a manner whereby the close interaction with the field influences the project outcome in a way where the outcome will be knowledge that is relevant to the local context in which the study is situated.

Explorative design and mixed methods

The aims and objectives of this study pointed toward an explorative design intended to obtain a constant ability to adjust to the realities studied and the increasing knowledge of the context. The study also employs a "mixed methods" approach as the overall methodological approach. The first data collection is qualitative in order to investigate the experience of the employers. Informed by the first data collection, quantitative data will be collected in order to test and refine the employees' experience. A qualitative data collection with the employees will then be conducted in order to understand the results of the quantitative study in detail and investigate *their* lived experience. The collected data is therefore not interpreted simultaneously, but rather in a sequential manner. By finishing the data collection with a focus group, it is aimed at validating the responses in the previous data sets.

By using mixed methods, it is aimed at producing knowledge that reflects the multi-dimensionality of the social realities, including the different dimensions of challenges regarding labor recruitment and retention in Greenland. Based on the philosophy of pragmatism, the social phenomenon is seen as multidimensional and therefore cannot be studied using any one method alone (Mason, 2006: 9f.).

The empirical basis for this study consists of four rounds of data collection, including semi-structured interviews with employer and employees alike, an employee survey, as well as focus group interviews. During the first round of data collection, Carol Bacchi's approach, "What's the problem represented to be?" (WPR) (Bacchi and Goodwin, 2016) will be employed as an analytical tool in order to analyze the companies' strategies and measures regarding recruitment and retention. By combining interviews with the WPR approach, the focus is on how the challenge is represented in the companies' measures and strategies, while the interviews with the employers investigate how they perceive the challenge. This sheds light on whether there is a gap between what companies do in practice versus how the employers perceive the problem. In general, the project aims at pointing to gaps between how the employers and employees perceive the challenge.

As the recruitment and retention of labor in Greenland is a broad and complex research problem, and very little data on the topic exists, the use of mixed methods is advantageous. Using either qualitative *or* quantitative data alone would only grasp parts of the problem. As mentioned earlier, reality, especially when studying human beings, is multidimensional. Using mixed methods helps to cope with the limitations in data and to create valid data, as the project is not restricted by only using one method. The methods used fed into each other in a manner whereby the data was able to speak and to test trends shown in one form of data collection against other data.

The topic is also perceived as sensitive among the actors and society in Greenland in general, as it is highly politicized. Using mixed methods helps to cope with potential distrust and the common assumptions about the topic, as it can provide multiple perspectives that can subsequently be critically interpreted against each other.

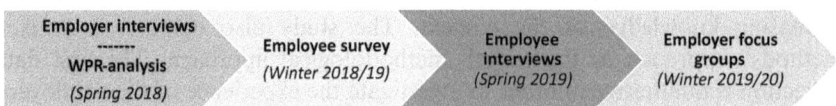

Figure 10.3 Preliminary data collection timeline

Thus far, after one round of qualitative data collection (employer inter-views) and the subsequent quantitative data collection (employee survey), the synergies between the methods provided insights into which trends are most interesting to investigate further in the third stage of data collection: the qualitative employee interviews. In order to let the data speak, the first qualitative data collection had a rather open structure, intended to allow the employers to mention the topics that they mainly associate with the recruiting-and-retaining challenge. The subsequent employee survey was structured by the findings from the first data collection but has also been open to other topics deemed relevant. The employee interviews, which at the time of writing have yet to be conducted, will then mainly be based on the survey findings, together with findings from the employer interviews. The final focus group interviews will be used to validate and discuss the overall findings.

As part of the research strategy, the data collection (see Figure 10.3) thus stretches from a rather open approach to an increasingly limited approach in order to let the data speak as much as possible and at the same time identify the most prominent findings.

Data collection

The data collection is done with three companies and two employee groups in each company. While the study is not a case study, the recruit-ment and retention challenges in Greenland are exemplified by the three chosen employers. Flyvbjerg (2001) argues for the power of a good exam-ple, even when traditional generalization cannot be achieved. He argues that in the study of human affairs especially, outcomes are not predictable, and the creation of concrete context-dependent knowledge is therefore more valuable than the quest for generalizations (Flyvbjerg, 2006: 73ff.). By including more than one example, the data in this project presents solid data with a large sample. The total number of informants will be approximately 1,200 employees. As shown in Figure 10.3, the data collec-tion is not conducted simultaneously; rather, it is accumulated in a sequential manner, starting with collecting data from the employers, pro-ceeding to data collection with the employees, and ultimately again dis-cussing the collected data with the employers to shed light on potential gaps in how they perceive the challenge. In PR, as in this study, the broad involvement of informants is important to avoid one sidedness and to obtain a certain level of representativeness. Moreover, even though tradi-tional generalizations cannot be achieved, the large sample will allow context-dependent generalizations.

The local partners are involved in each step of the data collection; either as an active part or in providing feedback, additional information, and more. This dynamic way of seeing the data collection is also an integral part of the explorative design and the locally anchored structure. The continuous interaction and dialogue with the local partners correspond to the applied phronetics, as the researcher grows close to the realities studied without becoming a subject in their own study (as for example in action research).

Being based on an exploratory approach, the aim was to involve all of the partners in the decision on which specific data material was to be collected (e.g., which employee groups should be in focus). Thus, the final six employee groups upon which the study is based were selected in close dialogue with the partners. This resulted in data collection that is not only in the interest of research but also in the interest of the company, as well as internally anchoring the project. The companies assisted in the internal communication of their involvement in the project, which facilitated the access to the employees. The close collaboration with the partners when choosing the employee groups can potentially be problematic, but this was not experienced as such in this project. The partners' knowledge of the field helped to qualify the overarching groups. Nevertheless, some frames still had to be matched, such as accessibility and group size.

As already mentioned briefly, the data collection is planned as a four-stage process in which the companies are continuously involved. They are to be presented with the preliminary findings and encouraged to give feedback. This feedback will be critically revised, and, if deemed relevant, will be implemented in the next phase of data collection. This means that the data is kept in the hands of the researcher, and the involved partners are presented only with selected data sets.

Informed consent was/will be obtained from the informants. In the first round of interviews (most of which were conducted via telephone), consent was given orally, whereas written consent will be requested in the subsequent data collection. Measures are also taken to assure confidentiality. Informants are offered anonymity regarding their name, specific position, and company.

In addition to the collected data as presented in Figure 10.3, observations and experiences in the fieldwork and everyday life while living in Greenland were made. The latter is particularly valuable with respect to the contextual local knowledge developed by the researcher in the course of the project. Informal informants who reached out to the researcher, observations from everyday life, as well as local media certainly had an influence on the researcher's understanding of the context. At the same time, by moving to Nuuk, the researcher herself became a subject in the context.

Clear expectations, but also commitment

Clear expectations toward each other's commitment to the project have been set with all of the partners. Apart from the project financing, the commitment comes from both sides: on the one side is the partners' willingness to invest their time in assisting me in data collection, internal knowledge dissemination, and answering general and specific questions; on the other side is the researcher's willingness to keep the partners updated throughout the process, to share preliminary results with them, and to become involved in shorter, relevant projects for Category 1 partners. Fixed time periods for these projects were agreed in each case, for example, assisting/working on a welcome package for new employees together with a human resources employee for one day a week over an eight-week period. The clarification of expectations toward one another also includes short evaluations on the project from time to time.

The latter commitment of working on shorter projects for Category 1 partners has proven to be especially beneficial. Wadel (1991) argues that it is not the role of the researcher to give access to the field, but that the researcher must put this role aside in order to gain access (Wadel, 1991, p. 27). He further argues that fieldwork should include some regular work, where the researcher is assigned another role (ibid.: 30).

Knowledge dissemination of the research and research results is a goal of PR and was requested of this project by the Category 1 partners. While informing the Category 1 partners during the research progress is a commitment from my side, all of the partners also offer assistance in mediating the project internally and externally at a later stage.

The project financing had been secured from the outset, including an extended grant (provided by the Category 1 partners) to cover the travel costs, which were expected to be higher than usual due to the high costs of travelling within Greenland. One of the external Category 1 partners has also been assigned the role as project leader and thus serves as the link between the partners and Aalborg University, which is my official employer. By delegating the tasks in this manner, the communication between the partners regarding the financing of the project has been smooth, as there was only one link, which helped me to be able to maintain my focus on my research.

For a project like this, one can argue that the need for relations and network in the local society is generally valid. As this project is built around actors who were part of the researcher's network before starting the current project, it is argued that their financial involvement in the project makes no significant difference to the content of the project; the only difference might be found in the researcher's active participation in some of the companies' internal activities during the research project. However, this commitment was also agreed on for the researcher to gain deeper knowledge of the organization and internal processes of the individual company.

Experiences in practice

As described in the previous section, the expectations of this collaboration between business, education, and research were aligned from the beginning. Still, my role as researcher in the collaboration presented advantages and challenges.

My own academic and professional background trained me in the tradition of research that is to the benefit of society; research generally involving local communities. As mentioned earlier, this type of research is neither equally suited to all problems nor to all researchers, but the specific research topic made an activating and collaborative research design necessary.

The everyday time and partner management has been the most demanding aspect of the project thus far. Apart from the officially involved partners, other parties continuously display interest in the project. I am occasionally contacted about new project ideas within the existing project or outside of it. In order to interpret a certain representation, the research must have prior knowledge, also called "collateral knowledge" (Peirce, 1931; Pennington and Thomsen, 2010). The interest in the study from the outside is primarily positive, as it contributes to my own learning process and understanding of the field. Informal talks and meetings with different informants enrich the data collection and ease my understanding of the context. The metadata thus adds to my own knowledge and reflections on the topic.

However, I also experience this as a balance between fulfilling the commitments to the partners at the same time as not actively turning down the interests of others, as their knowledge of the field remains very important, especially for networking. Usually, I would meet the latter over a cup of coffee to have an informal chat about the project and how they experience the challenges as a way of staying in touch with them. In some cases, it was necessary to make it particularly clear (as politely as possible) that it was not possible to add partners to the project and to refer the interested party to the final results. To avoid conflicts of interest, I always try to communicate clearly and honestly about my expectations and boundaries.

Topic-wise, the continuous input from different partners resulted in a flow of new ideas and project angles, especially early in the project. Again, this can be seen as an important part of the learning process, although it also required extensive effort on my part to avoid losing focus. Especially in the first project period, I spent considerable time and energy training my objectivity and trying not to get caught up in the subjective perspective of the partners. Based on this background and the aforementioned metadata, the research questions were continuously adjusted early on in order to explore different perspectives on the topic. At a certain point, it needed to become relatively fixed in order to maintain the focus and integrity of the research. This also accounts for maintaining one's integrity as researcher. In a way, this presents a double integrity: On the one hand, the research integrity of remaining independent and keeping one's integrity despite the continuous external input; on the other hand, my individual integrity as I myself became a subject in my field of research.

The above implies that it has often been necessary to invest considerable time in communication with the partners. In the initial phase of the project, this appeared to be advantageous; as the project has advanced, however, the focus has shifted toward more balanced and structured communication and interaction with the partners. The first months of the project period were used to take in and learn as much as possible about the context—even sometimes getting lost in the context. In my experience, this period presented an important learning period with respect to becoming aware of my own subjectivity while at the same time I gained in experience in how to stay objective and aim for objectivity. Throughout the project, I have experienced periods with close contact to partners and society and other periods with much less contact in order to focus on the project content. Kofoed and Staunæs (2015) argue for "hesitancy as an ethical act," claiming that:

> temporality plays a key part in hesitancy as a responsible ethical strategy. Hesitancy requires time, the opportunity to postpone action and a chance to "put the situation on hold." We argue that the pace with which we can create new knowledge and new insights is qualified by such pauses.
>
> (Kofoed and Staunæs, 2015: 37)

In order to settle my thoughts and reflect on the knowledge obtained through close contact with the field, I needed to take pauses from the field to the extent that this was possible (as I myself had become a subject within it). Thus far, in different periods of the project, I have had different needs regarding the contact to and communication with the field; needs that have had to be balanced. Even though the project was based from the outset on the researcher filling different roles at different times, I sometimes experienced this shift between different roles as rather demanding.

Another experience concerns the building of trust via the project's participatory method. As discussed earlier, collaboration with local partners is often regarded as a trust-building approach, which has also proven to be the case here. I sometimes experience distrust from "outsiders," however, usually either relating to my status as a foreigner or the local financing of the project. The distrust of me for not being a Greenlandic scholar could often be minimized by explaining the project, my own commitment toward it, and mention of the fact that I have actually formally transferred my residence to Greenland. In some instances, the fact that I am also trying to learn Greenlandic also made an impression on critics.

The financing-related skepticism is somewhat understandable, as the question of how I could maintain my independence as a researcher while the project is partly financed by local organizations was occasionally voiced. Again, as mentioned and discussed earlier, I attempted to overcome such concerns by describing and explaining the critical approach of the project, the use of mixed methods, and the project's approximation of objectivity.

Parallel to my close collaboration with local partners in the course of the project, I experienced periods in which I felt a lack of proper attachment to academia, especially in my everyday life. I was continuously in contact with my home university via online tools and planned my travels to Aalborg to be able to meet colleagues at research group meetings or similar activities. However, I was isolated from everyday methodological and theoretic discussions with my colleagues. My participation in seminars, courses, and workshops was limited. While I established good relations and contacts at the University of Greenland and had opportunity to participate in courses locally, it did not fully compensate for the academic environment that I was missing in Aalborg.

Finally, I would like to add some reflections on my role as an external researcher in Greenlandic society thus far.

When I first decided to move to Greenland for the duration of this study, the decision was mostly based on my own pragmatic understanding of the necessity to get closer to the realities under investigation. As I saw it, the study would be greatly enhanced by getting closer to the field. In retrospect, the decision to move to Nuuk with my partner provided much more than just extended access to the field; not least in terms of the creation of a deeper level of trust between the local partners and me, as they respected my decision to move here to conduct the study, despite the disadvantages and "administrative costs" resulting from my employer being located in Denmark. My experience has been that the choice to move to Greenland for an extended period has set me apart from hit-and-run researchers, and that this has been advantageous for my project.

By moving to Nuuk I gained extended access to the field and developed trust with those with whom I talk to during my investigation. Being a resident of Nuuk provides flexibility when making appointments, another type of presence in society than if I was merely visiting, as well as a different basis for discussion due to my extended contextual knowledge. I have also found that it has eased my data collection, as I can reach out more easily to locals (and also be reached out to). Some of these aspects can also be achieved by researchers who visit a location repeatedly or for long periods of time, as is usually the case with ethnographic researchers. Personally, I have found that actually residing in Nuuk provides important justification for conducting the project. The experience of living in Greenland—and (partly) becoming a member of the labor force myself—justifies my research on the Greenland labor market, especially compared to hit-and-run researchers.

I have also found my personal background to be advantageous when talking to locals. Being German and having only lived in Denmark for a few years has also been a factor. Being neither Greenlander nor Dane, I have been able to position myself outside of the potential ethno-cultural or post-colonial tensions, depending on the situation. Together with the fact that I have my main residence in Greenland, I experienced situations in which the atmosphere changed for the better when informants learned of my personal background. According to Rowe (2014), researchers should only reflexively discuss field experiences to the extent that it actually makes an impact on the research. The reaction to my presence as a researcher during my fieldwork varied (Rowe, 2014). Depending on the situation, I could be a researcher from Nuuk, a Danish researcher, a German researcher, a researcher financed by both Danish and Greenlandic organizations, and more. I avoided going into detail with my personal background when doing so was unnecessary. In situations where I experienced a sense of distrust or skepticism toward my role as a researcher, I found a way to briefly mention my personal background, which I have almost always experienced as making a positive difference. Drawing on my personal identity, I could often place myself outside of the existing discourses pertaining both to hit-and-run researchers, as well as Danish researchers in Greenland, which I experienced to be an advantage.

Especially in the context of smaller projects for Category 1 partners in my project, I have occasionally experienced being an external researcher in the organization as challenging. My role in the organization has been unclear for some: one day I could be a researcher; another day a "normal" employee assisting with daily tasks; and on a third day some kind of consultant, working on special tasks that are close to my research topic without actually being part of the research. Even though the division of tasks was clear both for me and my main contact in the company, it was not always clear for the other employees. I experienced becoming part of the department as challenging; in some cases, I felt like a stranger for some time, being somewhere between being a genuine colleague and an external visitor. This required constant communication with my colleagues and the head of the department, as the company was not accustomed to such projects. I have found that it has helped to be open to questions and proactive in my contact to my colleagues, as well as trying to be visible in the workplace. This requires extensive effort of me personally, and there have been times where I actively chose my home office instead of the workplace (see also the section above on how to balance research integrity).

However, the benefits of being an external researcher in Greenland outweigh the described challenges—my initial goal was extended access to the field and learning about the field, which has largely been a success thus far. To this can be added unanticipated benefits, such as the interest of local, informal informants.

Conclusion

Even though local collaboration in research projects has become fairly common, much remains to be learned about the research and "business" ends of the relationship. My PhD project shows how the collaboration is strongly based on the relations established with partners, together with consistent communications throughout the process. Each research collaboration is different, and participation-based research can have different dimensions. In my case, the participatory-based research design required the use of methods that allow the adjustment of the research strategy throughout the process, as close collaboration is expected to provide results that could require changes throughout the duration of the project. My use of mixed methods has enabled me to "allow the data to speak" while at the same time being the researcher steering the project.

Having paying partners in the project *can* make a difference to the project design. In my case, however, as the topic and initial approach to the project were designed in close collaboration with the partners and concern a topic of daily relevance for the companies, I did not (yet) experience constraints on the project resulting from this. Open communications and the clarification of expectations throughout the process have played an important role in this regard.

My main benefit was that my project investigates a topic of great relevance and significance for Greenland and thus has the ultimate goal of being of benefit to society. According to my experience, this has not only made it easier to get in contact and develop trust with the local partners; it has also contributed to the development of general interest in the society for the research. And conversely, the general interest in society contributed to the understanding of the research context.

Being present in the field has been of great advantage to my research, even though it entailed sacrifices in other areas, such as my everyday contact with academic colleagues.

The most important lessons learned thus far from engaging in PR in Greenland have been to be patient (creating trust and access to locals takes time), to be social (relations and network are crucial), to be present (in the field, for partners, etc.), to be open (to input, knowledge, contacts, etc.), and, finally, to balance the latter with research integrity by being outspoken, critical, and reflective about the different roles a researcher can have during PR.

As a result, my approach to participatory research provided particular knowledge and experience in the field that is not only expected to contribute to the relevance of the research and applicability in Greenland but can also serve as an inspiration for more locally anchored research internationally. While these lessons are collected based on specific experiences in Greenland, they are seen as being of great relevance for participatory research in similar environments.

Notes

1 "Hit-and-run" research: Where the researcher collects data locally and leaves immediately. Such research results might not be shared with the locals afterwards and are often talked about in a negative, critical tone in the context of research in Greenland, based on the assumption that the research results are often neither mediated to nor come to benefit the local society.

2 Arctic Consensus is a platform for more and better cooperation in the North Atlantic region and is based in Aalborg and Nuuk. Source: www.arctic-consensus. com.

3 For example: Presentation at the Arctic House, Aalborg, August 2016; Presentation for the Greenlandic members at the Danish Parliament, March 2017; Interview with KNR, Greenlandic Broadcast Company, Spring 2017; Presentation at Future Greenland Conference, May 2017.

4 Action research is concerned with both reflection and action, a process whereby the researcher actively participates in the implementation of solutions to a problem in the research process (see also Reason and Bradbury, 2006).

References

Bacchi, C. and Goodwin, S. (2016). *Poststructural Policy Analysis: A Guide to Practice*. (C. Bacchi, Ed.). New York: Palgrave Macmillan US.

Flyvbjerg, B. (2001). *Making Social Science Matter. Why Social Inquiry Fails and How it Can Succeed Again*. Cambridge: Cambridge University Press.

Flyvbjerg, B. (2006). Five misunderstandings about case-study research. *Qualitative Inquiry*, 12(2): 219–245. https://doi.org/10.1177/1077800405284363.

Gibbons, M. (1999). Science's new social contract with society. *Nature*, 402(6761supp): C81–C84. https://doi.org/10.1038/35011576.

Greene, J. C. (2007). *Mixed Methods in Social Inquiry*. San Francisco, CA: Jossey-Bass.

Greenland Research Council. (n.d.). Research. Retrieved from: http://nis.gl/en/research/.

Guthrie, G. (2010). *Basic Research Methods: An Entry to Social Science*. Delhi: SAGE Publications.

Hall, B. L. (1999). Looking back, Looking Forward—Reflections on the International Participatory Research Network. *Forests, Trees and People Newsletter*: 33–36.

Huppert, V. (2016). Brain Drain in the Arctic. The retention of Local High-Skilled Labour in Nuuk Greenland. Aalborg University.

Kofoed, J. and Staunæs, D. (2015). Hesitancy as ethics. *Reconceptualizing Educational Research Methodology*, 6(2). https://doi.org/10.7577/rerm.1559.

Kørnøv, L., Lyhne, I., Larsen, S. V., and Hansen, A. M. (2011). Change Agents in the Field of Strategic Environmental Assessment: What Does It Involve and What Potentials Does It Have for Research and Practice? *Journal of Environmental Assessment Policy & Management*, 13(2): 203–228. https://doi.org/10.1142/S1464333211003857.

Lofland, J. and Lofland, L. H. (1984). *Analyzing Social Settings: A Guide to Qualitative Observation and Analysis* (2nd ed.). Belmont, CA: Thomson Wadsworth.

Mason, J. (2006). Six strategies for mixing methods and linking data in social science research. *NCRM Working Paper Series, 2007*, 14. Retrieved from: http://eprints.ncrm.ac.uk/482/.

Morgan, D. L. (2014). Pragmatism as a Paradigm for Social Research. *Qualitative Inquiry*, 20(8): 1045–1053. https://doi.org/10.1177/1077800413513733.

Peirce, C. S. (1931). The Collected Papers of Charles Sanders Peirce. Retrieved from: https://doi.org/10.1038/135131a0.

Pennington, J. W. and Thomsen, R. C. (2010). A semiotic model of destination representations applied to cultural and heritage tourism marketing. *Scandinavian Journal of Hospitality and Tourism*, 10(1): 33–53 https://doi.org/10.1080/15022250903561895.

Reason, P. and Bradbury, H. (2006). *The SAGE Handbook of Action Research: Participative Inquiry and Practice London*. London: SAGE Publications.

Rowe, A. (2014). Situating the Self in Prison Research: Power, Identity, and Epistemology. *Qualitative Inquiry*, 20(4). Retrieved from: https://doi.org/10.1177/1077800413515830.

Wadel, C. (1991). *Feltarbeid i egen kultur. En innføring i kvalitativt orienteret samfunnsforskning*. Flekkefjord: SEEK.

11 Co-creating knowledge for and with the Arctic

Future avenues

Carina Ren & Anne Merrild Hansen

> We need a different approach. The 1800s will be remembered as a century of great Arctic explorers. And one of greater colonization. The 1900s will be remembered as a century of great scientific investigations. Both centuries failed, however, to recognize the value of the indigenous peoples of the Arctic. Let's do little things together which will help make this current century one of connections, one of working collectively in which the scientific community doesn't just come to Greenland or other parts of the Arctic to undertake research that is only of interest to it.
>
> (Lynge, 2007: x, quoted from Holm et al., 2011)

At the time of writing, registration to the first edition of the Greenland Science Week 2019 had just closed. According to the event's Facebook site, more than 200 delegates from 12 countries had registered for the main event, the Polar Research Day, held for the first time outside of Denmark. The event is illustrative of a general movement inaugurated in 2009 under the process of Greenlandic self-government, according to which more and more areas are "taken home" (*hjemtagne*) from Denmark. The event also reflects a steadily growing number of voices in Arctic societies as well as in scientific communities calling for an improved local grounding of knowledge. This includes the involvement of local stakeholders and communities in defining research scopes and aims, drawing on—and drawing in—local and indigenous knowledge in the research process and sharing research findings and outcomes with impacted communities.

As illustrated by the Polar Research Day and as suggested in the extract above, it is time for research to readdress and rethink the role and authority of local and indigenous voices and knowledge in research. It is also time for researchers from outside Greenland to reconsider their role, value(s), and relationships with local stakeholders, communities, politicians, and other researchers. This edited volume is but one contribution, one "little thing" among many others already unfolding and to come, to the ongoing conversation of how to connect and to work collectively around concerns of mutual interest and benefit to researchers, the inhabitants of Greenland, and the Arctic more generally. By sharing the experiences with different ways of designing, undertaking, and reflecting on collaborative research, we hope that this book will serve to inspire others to consider the need for collaborative Arctic research, together with the opportunities and benefits resulting from such collaboration. The contributions to this book have presented the reader with an array of recent and ongoing research projects in which researchers from AAU Arctic commit to collaborating in various ways with members or groups in local communities, municipalities, companies, and other relevant stakeholders in Greenland.

As laid out in Chapter 1, the authors do not commit to collaboration only because of the changing demands and requirements for involvement, representation, and participation, but also because of AAUs long-standing commitment to problem-based learning. Through problem-based learning, projects engage with and incorporate locally identified issues and challenges into their research, ideally from the early shaping of research questions and design to stakeholder integration in the research process and the iterative sharing and reworking of findings along the way. So although not all the contributors follow a cookie-cutter approach to designing collaborative research (or perhaps did not even see themselves as undertaking just that), their research is better equipped to address and promote locally desired social change through collaborative practices.

Research informs the decisions affecting the futures of Arctic communities. Owing to its ability to relate better than "hit-and-run" research, collaborative research is therefore of central value to local communities to local concerns. By way of example of how to bring new voices to the fore in research, our hope is that the approaches explored in this collection will inspire researchers to become even more attentive to local ownership, concerns, and impacts when framing, conducting, and communicating their research. This final chapter sums up some of the key findings arising across the chapters, discusses learning outcomes to help to prepare for future collaborative research projects, and reflects on what the future holds for external researchers in relation to collaborative research in Greenland.

Collaboration approaches: more than one, less than many

The first part of the book introduced different traditions in Arctic research and described collaborative research as gaining traction with researchers throughout the Arctic. Building on the personal accounts of researchers with close connections and working relations in Greenland, this changing reality is reflected and fleshed out in further detail in Chapters 2–4. These chapters display very different ways in which collaboration in and around research projects can be practiced and experienced. In conversation with Anne Merrild Hansen in Chapter 2, Minik Rosing discusses the ideological changes in the role of science in Greenland and the Arctic of today in the context of the Greenland Perspective project. In Chapter 3, Graugaard reflects on research subject positions in post-colonial contexts, while Andersen unravels his research practices in Chapter 4, describing the relations to practitioners in the field of health care in Greenland and the significant benefits this has had for research results and application. As the contributions illustrate, "collaborative inclinations"—whether related to general societal changes or shifting concerns of research positionality—each in their own way enable the production of new kinds of knowledge and understanding not warranted by or stemming from "hit-and-run research," as suggested in the foreword by AAU Arctic Chair Henrik Halkier.

The three introductory testimonies illustrate how the commitment to collaboration can assume various forms, and the practice of what is often referred to as community-based participatory research covers a range of action-oriented intentions, participation, and collaborative partnerships. In that sense, collaborative research can be said to be *more than one* but *less than many*. By this, we point to how collaborative research might be designed and carried out in various ways but should somehow be characterized by collaborative efforts and united by a sensibility towards making science matter and contributing to ongoing Arctic conversations. This is further illustrated in the second part of the book, Chapters 5–8, which introduce an array of cases based on research projects in Greenland.

As we reflect on the third and final section of the book, collaborations in contemporary universities are not limited to research initiatives. As two exemplary cases demonstrate, collaborative approaches are now woven into central fields of interaction in the Arctic research of today, here with respect to *education* (Chapter 9) and *knowledge collaboration* with external partners (Chapter 10). These chapters underline how collaboration neither is, nor should be reduced to research activity, but that it extends to areas of teaching and learning, as well as innovation.

The various experiences presented by the contributors to this book clearly illustrate how there is no one-size-fits all model when it comes to designing and carrying out collaborative research in Greenland. Rather, the applied methods suit the purpose of engagement in the specific project and local contexts. Variables include issues such as the technology used, the task required by and of locals, the issues for which participation is requested, the environment of the participant, how many people have been engaged, and through which channels they collaborate. The role of each partner in the project partnership also varies, based on project needs, goals, skills, and access to resources.

The question now becomes: What are the gaps in the current practices and what kind of "next steps" have unfolded in the process of working with this book. In other words, what has the scrutinizing of how we are conducting collaborative research in Greenland taught us? Despite the variety characterizing the projects presented in this book, we do identify a number of similar concerns and issues: *the politics of collaboration, supporting multiple ways of knowing* (including indigenous), *the building of local research capacity*, and *the future routes for Arctic collaborative research*. We reflect on these issues in the following section, which we see as central in advancing the agenda and strengthening the ambitions of future Arctic research.

The politics of collaboration and collaborators

The first recurrent concern in this edited volume focuses on the "collaborators" in our research projects. Collaborative research projects are fundamentally enabled and defined by the relationship between one or more stakeholders and/or communities and a researcher or team of researchers. The contributing authors all emphasize initial reflections and considerations with regards to identifying and defining the main stakeholders in relation to the individual project. Most also address challenges or worries about how or if they could or should be reached. For the researchers, collaboration with a community entails first and foremost defining what "the community" is.

Such a question must involve critical reflections about representation and about who gets to speak on behalf of the community on a certain case. We find an example in Chapter 6, where government officials, non-governmental organizations, companies, and researchers were engaged in the discussion of what and whom to include in a study and how it should be structured. Similarly, it is of great importance who gets to represent employers and employees in Greenland, as seen in Chapter 9, where Huppert presents an investigation of the recruitment and retention of labor.

While it is important to reflect on what defines and delimits a community—and who gets to represent it—we must also recognize the diversity of Greenland. Today a growing majority of Greenlanders reside in larger towns, where people live a modern life according to Western standards. Yet there is also a Greenlandic everyday life that unfolds in very small and remote settlements, where people depend more directly on the surrounding natural and living resources. Such differences call for particular attention when developing project ideas and research designs. Cultural contexts and differences can influence communication structures and power relations between community members and call for locally adapted methods or ways of engagement in different settlements.

The degree of involvement can vary in terms of how and to what extent different communities are willing and able to collaborate in research. Their availability to participate could also be obstructed or at least influenced by remoteness, inadequate infrastructure, and simply the way of life, where hunters and fishermen are away from home for long periods. In that sense, not all community members have equal access to research or to sharing knowledge, or any access at all. Here, a gender issue is also prevalent if collaborative research is framed only around organizations, activities, and/or industries where men are dominant. As Suzy Basile argued at the Polar Research Day 2019 in Nuuk, the representation and knowledge of indigenous Arctic women have been overlooked up until today and must be better included, including in collaborative research settings.

These examples of how contextual settings impede research collaboration for individuals and communities illustrate the politics of collaboration: that access or opportunity are not evenly distributed among its subjects, but rather shaped and enabled along differences in, among others, connectivity and infrastructure, language, and gender. To allow for more open access to collaboration with Arctic communities, researchers call for more transparency and reflexivity. This also raises questions pertaining to positionality; that is, the notion that personal and institutional values, views, and location in time and space influence our understanding of the world (Rose, 1997).

Issues of positionality suggest that there is always a relationship between researcher and researched, and that this relationship involves how the researcher interprets the lives of others from their own perspective. Knowledge can therefore be considered the product of a specific position reflecting particular places and spaces, which places demands on the reflexivity of the researcher concerning their position and the implications for research.

We can identify different positions in relation to the field when considering the researchers contributing to this volume: at a *distance*, as (first) presented by Ren and Thomsen, through *participatory processes,* as suggested by Larsen and Hansen, and *action research* (Næs and Dreyer); through *iterative processes* (Ren and Thomsen); using *fieldwork* (Graugaard), or using more playful, innovative methods such as *diagramming* (Chimirri) or as "*embedded*" (Huppert). In all these cases, the research outcomes and ultimately, what we can know are shaped by these positions. This assists some insights while disabling others. While this is indeed characteristic of all social science research, we must remain cognizant of the performativity of knowledge production and the repercussions of "blind spots" for research and, importantly, the local communities in which and together with whom we are working.

Toward multiple ways of knowing

In the Arctic regions that are considered part of North America, the (lack of) inclusion of indigenous knowledge in research settings has long raised discussion and criticism. Already in 2013, Sejersen (2013: 33) pondered why—with few exceptions—this had never been an issue in Europe and Greenland. In the context of Greenland, Sowa (2013) has remarked how the nation-building process, the major landmarks of which were the achievement of home-rule in 1973 and self-rule in 2009, has enabled the Greenlandic Inuit to develop "the most advanced form of self-government" (p. 184). According to Strandsbjerg (2014), this has meant that "while indigenous identity is often seen as an alternative mode of political organization compared to statehood in international relations, the processes unfolding in Greenland indicate what happens when there is a transition from being indigenous to being nation-state" (p. 260).

To Sowa, this transitional stage has had consequences for Inuit self-identification. The underlying logic is that concerns for own independence and, hence, the opportunity to enter a stage as sovereign has gone hand in hand with a distancing from indigenous claims. According to Sowa (2013), this is "inherent in the global model of indigeneity that indigenes exist in a 'natural' and 'pre-modern' state in contrast to 'enlightened,' 'modern' cultures that have founded their own independent sovereign states" (p. 185). This identified lack of an issue is reflected in this volume, where—with Graugaard's contribution as the most notable exception—indigenous identity and knowledge plays a very limited role.

But are times changing? In many contexts, we are seeing the re-emergence and reclaiming of indigenous ways of being and knowing—both "from-below" and "top-down." Indigenous ontologies are entering, as we saw, the stage of the Polar Research Day in Nuuk in 2019, the courtrooms in pending international cases on the whaling and sealskin trade, the funding calls for Arctic research proposals, as well as in the climate-change discourse (Thisted, 2013) and in tourism and branding initiatives (Ren et al., 2019). In all of these social spheres, indigenous and local cultures and identities are being proposed as alternatives to the Western norm.

One way to proceed is the strengthening of indigenous ways of knowing in the guidelines and a "toolkit" for "Community Based Participatory Research in Greenland" (Rink and Reimer, 2019: 7). This newly published document tells of how:

> Despite these differences in view points, the overwhelming majority of the population are of Inuit or mixed Inuit-Danish descent. In Greenland, Greenlanders and Danish citizens hold many professional and leadership roles in Greenland. As CBPR methodology works to de-colonize systems and balance power it is essential that researchers and community partners are clear about the difference between indigenous members of Greenlandic communities and Danish members of the community. This means being intentional in the engagement of Greenlandic community members as community advisory board members, tailoring research protocols to Greenlandic worldviews and belief systems, and gaining the Greenlandic perspective. This does not mean exclusion of Danish community members, or Greenlanders who are both Danish and Greenlandic, rather it requires commitment to building on the voices and identity of the indigenous Greenlanders.

As the above text calls for new standards, greater introspection, and a more critical societal stance with respect to research and the ability of research to speak universal truths, the toolkit provides a checklist for the planning of research projects to inspire and promote increased research–community collaboration in the future.

While most research discusses fault lines and sharp contrasts between Western and indigenous ways of being and knowing, researchers such as Kirsten Thisted point to Greenland's opportunity to challenge the notion of "fatal impact," according to which indigenous communities are doomed in the encounter with the White Man. According to Thisted (2013), "Young Greenlanders, not least, have long felt uncomfortable with the roles of victims that has [*sic.*] been assigned to them in the discourses of post-colonialism and indigenous peoples" (p. 231). In many of her analyses of artwork, music, TV series, and exhibitions, Thisted analyzes the hybrid performances of Greenlandic cultural identity informed by a vibrant, contested mix of pop culture, historical and colonial stereotypes, and current aspirations for independence.

This approach indicates that while Greenlandic identity is contested (both within the country and the international context), it is also skillfully maneuvering between essentialist positions in often-hybrid configurations (Ren, Gad, and Bjørst, 2019). Expectations and demands at all levels will surely shape the co-creation of more diverse ways of knowing for years to come, which will allow for increased sensitivity and the inclusion of indigenous knowledge (or more broadly speaking: local knowledge). In this work, the hybrid experiences from Greenland might be highly relevant to inform the work to strengthen the existing or developing new ways of situated and collaborative knowing in the Arctic.

Building local research capacity

The original aim of this book was to introduce collaborative Arctic research and to exemplify this approach using a number of case studies carried out in Greenland by members of the Arctic research network at Aalborg University. While most of the contributors proved to have many long-standing and close ties and relationships to Greenlandic society, few of them are actually Greenlandic or long-time residents of Greenland. This reflects a situation where, until recently, research in and about Greenland has been driven by Danish research institutions and universities and by researchers such as (most of) us, who live south of the Arctic Circle. Such research has informed decisions on investments and policies about the development of infrastructure, social initiatives, resource exploitation, and fisheries, to name just a few of the affected areas.

While the intention of the book was never to present *Greenlandic* research, its composition of researcher profiles does address the growing aspirations for research that not only is conducted *in* and, increasingly *for*, the Arctic, but also research that includes—and is preferably led by—Arctic scholars and researchers. One might provocatively (yet fairly) ask how much the previously asymmetrical circulation of knowledge and ultimately power have really changed in Danish post-colonial research projects and practices in Greenland, when the capacity to initiate, frame, and undertake research still lies in the hands of the former colonizers.

Fortunately, this image misrepresents the Greenland research landscape. Through the development of Asiaq, the Greenland Institute of Natural Resources, and most recently the University of Greenland, the Health Research Centre in Nuuk, and the Nuuk office of the Geological Survey of Denmark and Greenland, local research institutions are increasingly assuming a leading role in educating the Arctic researchers of tomorrow and in research activities and management.

Despite this, it is still broadly recognized that Greenland research institutions have limited access to resources and that collaboration in Arctic research (just like everywhere else) is needed to get the most out of the ongoing research activities. Emerging learning and research initiatives, for instance with Nordic Arctic research institutions and together with the neighboring Iceland and Faroe Islands, bear the promise not only of new and meaningful research collaborations but also of joint capacity-building and knowledge-sharing for the benefit of societies and communities in the North Atlantic region.

Rethinking future research relations

Interrelated climate and geopolitical changes in the Arctic have intensified the interest from old and new members of the global research collective. Researchers from around the world are interested in coming to Greenland, and the management of these activities in a manner that protects local interests could require increased attention. The above guide is one of the many possible tools for ensuring this. However, combining a solid local research base with international collaboration remains the gold standard for research. In the years to come, balancing these two poles will require hard work—but doing so also promises major opportunities for the Greenlandic research community.

While intensified international collaboration opens up promising avenues of research for scholars from Greenland and the Arctic, as mentioned in the previous section, it also heralds new days for Danish researchers and research institutions, who, in this post-colonial era, can no longer expect an invitation to join in on research in Greenland in the same manner as previously. As we look forward and venture into the "century of connections" (Lynge, 2007: x), joining the table will demand bringing something to it—something for everyone to share and appreciate. With this edited volume as a modest dish for this potluck, we hope to have whetted the reader's appetite to engage in collaborative research, hereby crafting collaborative ways of knowing the Arctic.

References

Holm, L., Grenoble, L., and Virginia, R. (2011). A praxis for ethical research and scientific conduct in Greenland. *Études/Inuit/Studies*, 35(1–2): 187–200. doi:10.7202/1012841.

Lynge, A. (2007). Arctic Warming at the Tipping Point: An Inuit Voice, lecture presented at Dartmouth College, February 27.

Ren, C., Gad, U. P., and Bjørst, L. R. (2019). Branding on the Nordic margins: Greenland brand configurations. In C. Cassinger, A. Lucarelli, and S. Gyimóthy. (eds.), *The Nordic Wave in Place Branding: Moving Back and Forth in Time and Space* (160–174). Cheltenham: Edward Elgar Publishing.

Rink, E., Reimer Adler, G., et al. (2019). *Community-based Participatory Research in Greenland*. Nuuk: Ilisimatusarfik.

Rose, G. (1997). Situating knowledges: Positionality, reflexivities and other tactics. *Progress in Human Geography*, 21(3): 305–320. doi:10.1191/030913297673302122.

Sejersen, F. (2004). Local knowledge in Greenland: Arctic perspectives and contextual differences. In D. G. Anderson and M. Nuttall (Eds.), *Cultivating Arctic Landscapes: Knowing and Managing Animals in the Circumpolar North* (33–56). New York: Berghahn Books.

Sejersen, F. (2013). The Indigenous Space and Marginalized Peoples in the United Nations. *Acta Borealia*, 30(2): 223–226. doi:10.1080/08003831.2013.843308.

Sowa, F. (2013). Relations of power and domination in a world polity: The politics of indigeneity and national identity in Greenland. *Arctic Yearbook*: 184–198.

Strandsbjerg, J. (2014). Making sense of contemporary Greenland: Indigeneity, resources and sovereignty. In R. C. Powell and K. Dodds, *Polar Geopolitics? Knowledges, Resources and Legal Regimes*: 259. Cheltenham: Edward Elgar Publishing.

Thisted, K. (2013). Discourses of indigeneity: Branding Greenland in the age of self-government and climate change. In S. Sörlin (Ed.), *Science, Geopolitics and Culture in the Polar Region: Norden Beyond Borders* (227–258). Abingdon: Routledge.

Index

Page numbers in italics refer to figures. Page numbers in bold refer to tables. Page numbers followed by "n" refer to notes.

For Product Safety Concerns and Information please contact our EU
representative GPSR@taylorandfrancis.com
Taylor & Francis Verlag GmbH, Kaufingerstraße 24, 80331 München, Germany